DEADLY
Departure

DEADLY
Departure

Why the Experts Failed to
Prevent the TWA Flight 800
Disaster and How It
Could Happen Again

CHRISTINE NEGRONI

Cliff Street Books

An Imprint of HarperCollins*Publishers*

DEADLY DEPARTURE. Copyright © 2000 by Christine Negroni. All rights reserved. Printed in the United States of America. No part of this book may be used or reproduced in any manner whatsoever without written permission except in the case of brief quotations embodied in critical articles and reviews. For information address HarperCollins Publishers Inc., 10 East 53rd Street, New York, NY 10022.

HarperCollins books may be purchased for educational, business, or sales promotional use. For information please write: Special Markets Department, HarperCollins Publishers Inc., 10 East 53rd Street, New York, NY 10022.

FIRST EDITION

Designed by William Ruoto

Printed on acid-free paper.

Library of Congress Cataloging-in-Publication Data has been applied for.

ISBN 0-06-019477-4

00 01 02 03 04 ❖/RRD 10 9 8 7 6 5 4 3 2 1

For my father

ACKNOWLEDGMENTS

Despite the horror, anger, frustration, and grief that followed the crash of Flight 800, hundreds of participants agreed to tell their personal stories. Without them *Deadly Departure* could not have been written.

Thanks to the Cremades Vila family, the Gray brothers, the Houck family, the Krick family, Ted Lang, Maria Lucien, and the Snyder brothers.

I owe the following people for the time and energy they generously shared:

Bill Adair, First Officer Vincent Cocca, Mark DiPalmo, Tom Ellis, Peter Goelz, Bob Golden, Kitty Higgins, Jim Kallstrom, Elaine Kamarck, Mike Kelly, Bob Knapp, Lee Kreindler, Lieutenant Thomas Martorano, George Marlin, Don Nolan, Captain Jerry Rekart, Pat Robinson, Captain John Rohlfing, Ron Schleede, Dr. Dennis Shanahan, Captain Robert Sumwalt, Dr. Charles Wetli, and Arthur Wolk.

In addition, technical advice and historical background were shared with infinite patience by Captain Fred Arenas, Captain Eugene Banning, Merritt Birky, Ed Block, John Borger, Benito Botteri, Ray Boushie, William Brookley, Robert Clodfelter,

Major General Richard Goetze, Richard Hill, Thomas Horeff, Michael Huhn, Cleve Kimmel, Craig Mullen, Jimmie Oxley, Ned Preston, Michael O'Rourke, Captain Hugh Schoelzel, Carl Vogt, Jim Wildey, Captain Jim Walters.

I am grateful to the following press representatives: Doug Webb, Gary Lesser and Russ Young at Boeing, David Venz and Mary Anne Greczyn at Airbus, Eliot Brenner and Kathryn Creedy at the Federal Aviation Administration, John Mazor at the Air Line Pilots Association, Mark Hess at the National Aeronautics and Space Administration, Jim Finkle at the Air National Guard 106th Air Rescue Wing, Ted Lopatkiewicz and Paul Schlamm at the National Transportation Safety Board. Thanks also to Sue Baker and Helen Cavanaugh at Wright Patterson Air Force Base, and the Federal Bureau of Investigation.

Special thanks go to my agent, Freya Manston, for believing in my work, and to my publisher, Diane Reverand, and Matthew Guma at Cliff Street Books.

I was challenged and encouraged by CNN producers Nancy Peckenham, Barclay Palmer, Ron Dunsky, and Sharon Troetel, and well served by the CNN crews who worked with me, made me laugh, and even fed me.

My sister Lee read every word of this book and cut out half of them. She's the best editor a sister can have. My children, Marian, Antonio, Sam, and Joseph, did without me for months on end without complaint and indulged me with their support and encouragement.

Everlasting love to my husband, Jim Schembari, who knows what's good, right, and important, and by his example teaches me to strive for all three.

PREFACE

We will never know what the last moments were like for the people who died on Flight 800. I have described those scenes by piecing together evidence from a number of sources, including transcripts from the cockpit voice recorder and conversations between the pilots and air traffic controllers; examination of cockpit instruments and the flight data recorder also provided important details.

From 8:31 P.M., the time the plane exploded without warning, there is less verifiable information about events. With the help of numerous experts I have written what is likely to have happened.

I spoke with fellow pilots and supervisors who worked with the crew. They were able to offer insight into their personalities, motivations, skills, and philosophy of piloting.

Pilots who lived through harrowing flights analyzed their experiences to construct a scenario of the last actions of Captains Steven Snyder and Ralph Kevorkian, and flight engineers Richard Campbell and Oliver Krick. Professional flight attendants drew detailed scenes of the atmosphere in the passenger cabin at the beginning of a transatlantic flight. They are well suited to do this because so much of their job requires observation of passenger behavior.

Aerodynamicists, who study the physics of flight; air crash investigators, who find answers in a confusing jumble of wreckage; and physicians in aerospace medicine, who study the effects of force on the human body, talked with me about what happened aboard TWA Flight 800 from the moment it exploded at 13,700 feet. Family and friends of those who died were eager to talk about the character of their loved ones and even, with some difficulty and pain, willing to consider how they may have faced death.

I became involved in the crash on July 17, 1996, the night it happened. Just before midnight, my husband shook me awake. He had heard on the news that a TWA 747 had crashed shortly after takeoff from John F. Kennedy International Airport in New York. He was bracing me for the call that was certain to come from CNN, where I worked as a correspondent. He was also bracing himself against his own fear of flying, which was brought very close to home with this particular crash. Days earlier, we'd buckled ourselves and our children into a TWA 747, returning from vacation in Europe. Snug in our seats, we were trustful but a little nervous. I am certain that many of the people who boarded TWA Flight 800 later that same month felt the same contradictory emotions: nervous about the improbability of such an enormous airplane safely flying anywhere, yet confident that those in control, the airline, the plane's manufacturer, the federal government, all knew what they were doing.

Before dawn the following day, I was in Long Island reporting the story for CNN. It was an assignment that would last more than a year.

Though the common perception of the Flight 800 disaster is that its cause will always remain a mystery, investigators discovered that the crash would not have occurred were it not for a fuel system design that has troubled aviation safety experts for years. Proposals to address the problem had been made, discussed, and dismissed for nearly four decades.

TWA Flight 800 was not the first airplane to be brought down by an in-flight fuel tank explosion. It was just the most widely publicized, the most dramatic, the most controversial. It was also the most deadly. That explains why until July 17, 1996, regulators and industry considered flying jetliners with flammable fuel tanks an acceptable risk, not worthy of a fix. The loss of 230 people convinced them otherwise.

AUTHOR NOTE

The comments that appear in quotation marks in *Deadly Departure* come from taped interviews I conducted with hundreds of individuals who were associated with TWA Flight 800, its crash and aftermath. Quotation marks also appear around transcripts of recorded cockpit conversations and communications between pilots, air control towers, and air controllers, as well as statements made to investigators, in public testimony, or legal depositions and statements made to news organizations. Where no verbatim record of private conversations exists, quotes were reconstructed from the recollections of the participants and substantiated. These reconstructed conversations will appear without quotation marks.

O God of earth and space. Of sea and fire and air,
Your providence surrounds us here and everywhere.

Jane Parker Huber, *A Singing Faith*

1

In a matter of seconds, the men in the cockpit realized they were going to die. In the minute that passed before the plane hit the water, fifty-seven-year-old pilot Steven Snyder was probably astonished that the Boeing 747, a plane he knew intimately and trusted completely, was failing him. Oliver Krick, the twenty-five-year-old flight engineer on the verge of becoming a commercial airline pilot, was likely feeling a different and unfamiliar emotion. For the first time in a life filled only with accomplishments, Oliver Krick felt helpless.

Thirteen minutes into the flight, the plane was still climbing out of New York airspace. There had been an explosion closely followed by a disorienting tempest of unrecognizable sounds. The force behind the noise shook the flight deck. When a quick fog of condensation filled the cockpit, the men grabbed for their oxygen masks and set the control knobs to the emergency position to begin a flow of pressurized oxygen.

Pilot training always includes time in a flight simulator practicing for in-flight emergencies, but there's no practice for the situation that was facing the pilots of TWA Flight 800. They did not know it, but the plane had split apart.

Desperate, Captain Snyder ordered flight engineer Krick to check essential power, looking for some reason why the battery in the electronics bay beneath the cockpit wasn't supplying an emergency source of energy for the flight control instruments. Krick was confused, unable to comprehend the sudden shift from normal to unimaginable. It might have crossed his mind that he'd done something wrong, and he was frantically reconstructing his actions.

Struggling against the cockpit's wild pitching, training flight engineer Richard Campbell eyed the panel by Krick, noting that the emergency battery switch was already in the "on" position. It should be providing electricity to the cockpit instruments. Yet dozens of amber flags had popped up in the flight control dials, indicating they were powerless. So was the crew.

On July 17, 1996, the Boeing 747-100 that was TWA's Flight 800 to Paris was one day younger than Oliver Krick. It had come off the assembly line in Everett, Washington, on July 15, 1971, the 153rd 747 made, and given the tail number N93119. Twenty-five is young for a man, but it's old for an airplane. Though this 747 still looked modern from the outside, its technology was essentially the same as that of the first 747 flown in 1969. The Boeing 747 and twin-engine 737 are the oldest commercial Boeing designs still in production.

N93119 had been scrupulously inspected by TWA under an FAA program to prevent age-related structural weaknesses, but the plane's systems were as old as the plane, including hundreds of miles of wiring that hadn't been examined since the day it was installed. The cockpit was a quaint array of yesterday's technology, dials and knobs, toggle switches, and analog gauges. There were no color graphical displays, no whiz-bang computers capable of improving on the calculations of the human flight engineers in a fraction of the time. This model

747, referred to as a "747 Classic," is one of the few remaining commercial jetliners still requiring a third crew member, like flight engineer Krick, to monitor the amount of fuel in the tanks and the operation of the engines.

In its twenty-five years, the jumbo jet had made 16,000 flights. It had flown 100,000 miles in just the last two weeks, making twenty-four transatlantic flights. It checked out fine as pilots Snyder and Ralph Kevorkian, and Campbell and Krick prepared it for the scheduled 7 P.M. departure to Paris. Reports filed by the pilots who'd brought the plane to New York from Athens showed nothing unusual during their nine-hour-and-forty-five-minute flight.

The ground crew at Kennedy noted that the Athens to New York leg had drained the fuel tank located between the wings down to the last fifty gallons, but since that tank would not be needed for the shorter trip to France, it was not refilled. Thirty thousand gallons of Jet A kerosene would be pumped into the plane's six wing tanks only. The wing tanks held enough fuel to get the plane to Paris: Filling the center tank would have increased the plane's weight, making the flight more expensive to operate.

TWA would have been pleased with more passengers. In the height of the summer vacation travel season, the 433-seat widebody was carrying only 176 fare-paying passengers. The fifty-four others on the flight were TWA employees and their families working the flight or enjoying free travel, the benefit of working for an airline.

Snyder, Kevorkian, Campbell, and Krick were not planning to fly Flight 800 to Paris. Their scheduled trip to Rome on TWA Flight 848 was canceled, so both passengers and crew were switched onto the Paris flight, which would continue to Rome after stopping in France.

Rather than go as passengers, a practice known as deadheading, as TWA schedulers had arranged, Krick and Captain Kevorkian were flying because Captain Snyder convinced New

York's chief pilot, Captain Hugh Schoelzel, to let them get the experience. Kevorkian would be completing his last supervised flight.

These fellas are on check rides, Hugh, Captain Snyder pleaded. Why not give us this trip and let 800's original crew deadhead into Paris?

And so it was that Oliver Krick, lucky from the day he was born, found himself in one of the best seats in the sky, two miles above the rustic shoreline of southern Long Island and climbing.

Phillip Yothers made the three-hour trip from central Pennsylvania to JFK Airport hundreds of times in his fifteen years driving for Susquehanna Trailways bus line. It wasn't his favorite assignment: New York traffic was always heavy, and sitting on a slow-moving highway, his right foot ping-ponging between the brake and the accelerator, his left leg riding the clutch, could make his muscles sore for a day or two.

The passengers on his bus on July 17, 1996, were high school French students headed for a week in Paris. He was drawn into their good humor and was soon participating in the banter of the kids sitting up front near his seat at the wheel of the big motorcoach.

Yothers, sixty-six, had never been to Paris, never traveled farther than he could drive. Even if he had the opportunity, he wasn't sure he would make such a trip. Yothers was the kind of man who liked to be home at night.

Earlier in the day, with the bus idling in the parking lot of Montoursville High School, he heaved luggage into the bins beneath the bus. Over the low rumble of the engine, and the high-energy chatter of teens starting an adventure, he'd overheard the French teacher tell a friend she really didn't want to go. Deborah Dickey and her husband, Douglas, were leaving two young daughters behind with their grandparents. Yothers, a father and a grandfather, understood her hesitation.

By the time the bus finally pulled up to the curb at the airport, Yothers noticed Deborah Dickey had caught the excitement. Students and their five chaperones burst out of the bus, snatching worn duffel bags and bulging backpacks from the luggage bins so fast that Yothers hardly had the chance to help. When he lowered the doors and turned to wave good-bye, every single one of them had already disappeared into the airport terminal.

If she hadn't been so eager to get back to Rome, Monica Omiccioli would have been excited by an unexpected trip to Paris. After all, Paris was fashion and fashion was Monica's other love. When she and her new husband, Mirco Buttaroni, arrived at Kennedy International Airport from Santo Domingo to learn their flight to Rome was canceled, it was just a frustrating delay for the honeymooning couple on their way home.

Monica was as colorful and dramatic as the clothing she designed in art school; she'd sketched a classic pinstripe suit rendered in scarlet, and fringed cowboy palazzo pants slit to the thigh. The twenty-five-year-old from Lucrezia, Italy, was always mixing things up in her work, in her life.

On June 23, 1996, she married Mirco Buttaroni, a banker she'd known since both were sixteen years old. Four days later they left for their three-week honeymoon. It was their first trip abroad, their first trip on an airplane. Devoted to the Catholic Church and to their large extended families, they planned to raise their own kids in the small village where they grew up.

Monica worked with her uncle at his design house, J Cab, in Fano, in northeastern Italy, producing men's fashions for an international market. Having graduated with honors in accounting from Luiss University in Rome, Mirco worked at a bank in the same town.

On their wedding day, guests took photographs of the beaming couple. Less than a month later, the snapshots were heartbreaking evidence of how quickly joy can turn to grief.

Boarding Flight 800, even novice flyers like Monica and Mirco had to give some thought to their vulnerability to terrorists. Security at Kennedy International Airport was heightened in the summer of 1996 because of the Federal trial of Ramzi Ahmed Yousef in a courtroom in Manhattan. Yousef was on trial for conspiring with notorious Saudi millionaire terrorist Osama bin Laden, to put bombs on U.S. air carriers flying into Asia. Passenger security checks and questioning by ticket agents is intended, in part, to help passengers recognize their role in keeping air travel safe.

But on July 17, no one was questioning the plane itself. The 747, an icon of the jet age, inspired confidence and awe.

As they prepared for the trip in the pilots' meeting room at Hangar 12, the airline's run-down operations building, TWA's economic troubles were very much on the minds of the pilots. Snyder was keen on figuring out how various operating procedures could reduce fuel consumption and help the airline get the most out of every gallon. His plan for saving fuel by making hourly adjustments to the plane's trim, its position in the sky, was referred to by some pilots as "Snyderizing" the plane. Since TWA's annual fuel budget comes close to $1 billion, a small savings adds up when multiplied by 300,000 flights a year.

At TWA, Snyder was considered the "godfather of the 747," an expert in the plane's fuel consumption. Some pilots wondered about his obsession with the subject, noting that he kept voluminous records detailing the fuel burn on each airplane he flew.

Seated on the flight deck, Snyder waited to depart, delayed because a computer could not match a checked suitcase to a passenger boarding card. Security precautions resulting from the 1988 Christmastime bombing of Pan Am 103 over Scotland forbade unaccompanied baggage on international flights. When the bag's owner was found on the plane and the suitcase reloaded in the cargo hold, the luggage loader broke down, blocking the 747 at the gate. More time passed before a tow truck could move the machine.

Restless passengers shifted around the cabin, finding better seats, looking for empty rows where they could bed down for the overnight flight. Some had been served drinks and were already using the bathrooms. Others were taking down their carry-on luggage, rummaging for an aspirin or a new CD. Getting things restowed and passengers belted back into their seats would not be quick or easy.

The flight was an hour behind schedule when the door to the cockpit opened.

"Hello darlin'," Krick drawled to the flight attendant who was giving him a thumbs-up.

"Everybody seated?" he asked, confirming her gesture. "Thanks."

"Amazing," Captain Kevorkian mumbled, relieved to be ready for takeoff.

Kevorkian, fifty-eight, was worried about how to explain the delay to passengers. There wasn't any good reason. The owner of the unidentified luggage had been on the plane all along.

"We won't bother telling them that," he said to the others on the flight deck. "You don't mind, huh?" he asked them, smiling.

Krick was working only his sixth flight with TWA, but he piped up, "We'd have a mutiny back there."

Had it been necessary to calm frustrated passengers, though, Krick would have been up to the job. Handling people was one of the things he did best. Flying planes and playing sports were the other two. He'd been competing in athletics since grade school. Water and snow skiing, soccer, hockey, golf, football, volleyball, basketball, even darts. In a boyish display, he'd wedged the plaques and trophies he won over the years into every inch of available space on the bookshelves in his bedroom. The only other decoration was an equally impressive collection of books, tapes, posters, and computer programs about flying.

Before joining TWA, Krick had been a flight instructor and corporate pilot and had recently been accepted in the Air National Guard for flight training on the F-15. He continued to live at home with his parents and younger brother in suburban St. Louis, but owned property on a lake in rural Missouri where he planned to build a house in the future. That future included Tiffany Gates, a woman he'd met in college and had dated steadily for five years. Ollie Krick was attractive, talented, and much loved. It occurred to him often how much he'd been blessed.

Charles Henry Gray III, the chief operating officer of the Midland Financial Group, missed his flight from Hartford to Washington's Dulles Airport, where he was scheduled to fly out to Paris, because his driver got lost on the way to the airport. He was rescheduled onto TWA Flight 800 along with his travel companions.

Later, as he waited in the airline's first-class lounge, his anger still in low idle, Gray called Elena Barham, the company's chief financial officer. "Damn, Ebie," he said, complaining about the screw-up in Hartford. Barham, who was also Gray's best friend, cheered him up, as she often did.

Gray had a thick shock of sandy brown hair, clear blue eyes, and a lopsided smile. He was tall and kept himself fit with daily five-mile runs. Raised in Arkansas, he developed a taste for good wine and well-made clothes once he left home. At forty-seven, he'd made and lost three fortunes and four marriages. There was something about his recklessness that, rather than alienating others, made him more endearing. He was a less-than-attentive parent. He didn't often see his two sons by his first and second wives when they were young. When Hank IV and Chad entered their teens and joined their dad in his perpetual adolescence, the relationship got going.

Gray had a custom twin-turbo Corvette that he would take to an uninterrupted five-mile stretch of Tennessee back road. With the boys in the car, he would run it up to 190 miles an hour and scream over the roar of the engine and the buffeting wind, "This is how I'm gonna go, boys, I'll die before I'm fifty and I'm gonna be goin' fast when I go." And though young Hank and Chad heard him say this many times, they never detected a hint of regret in their father's voice.

At six foot three, Gray had to duck his head slightly to board the plane. He settled in seat 2A in the first-class section. His mood had improved, courtesy of the champagne he'd been served in the TWA Constellation Club and the knowledge that this trip to Paris could make him very wealthy.

Gray was sitting in front of Kurt Rhein, with whom he was traveling to Europe to find financing for a merger of their two companies. A year earlier, Gray and Rhein had met on an airplane. The two hit it off right away. They developed a plan to merge Gray's specialized auto insurance company with Rhein's Danielson Holding Corporation.

Along with two other men on the flight, they were looking for the investors to make the idea a reality. Godi Notes, twenty-seven, an Israeli-born American who was an up-and-coming executive of the investment banking firm assisting in the merger, Donaldson, Lufkin and Jenrette, sat across the aisle from Rhein. Forty-one-year-old William Story, president of a Danielson subsidiary in California, was in one of the eight first-class seats directly behind the cockpit on the upper deck, the domed section that gives the 747 its distinctive appearance.

In front of Story sat Jed Johnson, forty-seven, who ran a New York–based interior decorating business with his partner and companion, Alan Wanzenberg. The two had established a golden reputation for designs that were tasteful yet experimental and a world away from Johnson's Minnesota roots. The designer's boy-next-door good looks were straight out of the sixties and suited the nineties infatuation with retro chic. Johnson was very much in demand as a decorator with a roster of celebrity clients. His work was often featured in glossy shelter magazines and books.

Johnson was alone on this trip, shopping for a new textiles business he was starting. Wanzenberg, whom he'd known for fifteen years, stayed in New York taking care of the businesses in Manhattan and Southampton.

Jed Johnson had a twin brother, Jay. He was one of four people on the plane who had a twin. The others were Arlene Johnsen, a TWA flight attendant who had lived with her sister, Marlene, all her life, and Myriam Bellazoug, a New York architect whose twin was Jasmine. Passenger Katrina Rose had a twin brother, John. None of the four was traveling with his or her twin.

Judith Yee had only one misgiving about this trip, which she'd been planning for months. She worried that her beloved terrier, Max, was going to be miserable. Before leaving for the airport, she coaxed him into the small pet carrier. It was wedged securely underneath the seat in front of her, and occasionally she could see his small nose through the grate of the plastic carrier.

Judy Yee was fifty-three years old, but not a strand of gray could be found in her shiny, chin-length black hair, which she was in the habit of pushing off her wide forehead with her hand, stopping along the way to nudge large black glasses back up on her nose.

Her dog was her companion and more than a pet; he was a coworker. For the past two years she and Max spent nearly every Thursday morning at P.S. 138, a public middle school in New York's Greenwich Village, providing severely disabled students with the opportunity to interact with the dog.

Judy and Max and another volunteer, Naomi Boak and her Labrador retriever, would arrive at the school first thing in the morning. Judy would place Max on a table where the children, some blind, some in wheelchairs, would gather around and pet or groom him.

Neither Judy nor Naomi had children of her own, so it wasn't surprising to either of them that they were developing an interest in the students at P.S. 138. Following their Thursday visits, they would walk together through the neighborhood, discussing the progress of the kids to whom they'd grown especially close.

As the 1996 school year drew to a close, the conversation also included Yee's plans to travel to France with a group of old friends. Yee had a take-charge personality and organized practically every detail of the trip. She coordinated each traveler's itinerary and handwrote the details of their trips on the pink pages of pretty travel journals she'd bound by hand for each of them.

In Yee, the women had a generous friend and their own personal travel agent. She found and secured the house in France where they'd be staying, organized the car rental, and booked the airline tickets. With characteristic devotion to Max, Yee chose TWA because it was the only airline that would allow her to keep the dog with her in the passenger cabin, rather than a pressurized cargo hold.

Judy Yee sat close to the front of the plane with her cousin from New Jersey, Patricia Loo. The third woman traveling with them, fifty-three-year-old Angela Murta, a beautiful blond with a wide open face and dramatic green eyes, had been assigned to a seat far back behind the wing on the right-hand side of the 747.

Friends of the women found the relationship between Yee and Murta a delightful contradiction. Murta loved clothes and boyfriends and scrimped on some things to afford an elegant house in the Hamptons. Yee, on the other hand, was quite well off, not terribly interested in men, and she spent extravagantly on only two things: art and travel. She dressed in sloppy, comfortable clothes that mirrored the way she kept her Greenwich Village apartment. Despite their differences, the women had been dear friends since the 1970s. Since so many seats were empty on TWA Flight 800, Murta could have moved up to sit closer to the women she'd known half her life.

As the passengers waited for takeoff, air handling packs beneath the plane had been running for nearly three hours, keeping the passenger cabin comfortably cool. Outside it was 71 degrees, a cloudless, moonless, nearly windless summer evening. A perfect night for flying.

3

With the uncertainty about who would be flying the trip to Paris and the last-minute delays at the gate, it had been a brittle few hours for the men in the cockpit. Not that the unexpected is unusual. The people-moving business is always subject to the unpredictability of passengers, the vagaries of weather, and the inconsistent reliability of extremely complicated machinery.

Shortly after 8 P.M., the paperwork for Flight 800 was completed, the plane readied, the last-minute hassles with the gate agents and luggage handlers resolved. The doors were securely closed and the pilots cleared to bring their 747 into a line of colorful giants, a United Nations of airliners into which Kevorkian could steer the red-and-white jumbo jet bearing the logo of a company that had pioneered commercial aviation.

Even if TWA's domination of the skies was on the wane, from where he sat, three stories above the airfield, Captain Kevorkian was at the top of his game. He was in command of the biggest and most awe-inspiring passenger jet ever made, operating out of one of the biggest and most glamorous airports in the world. For most of the upcoming flight, he would fly the 747 on an isolated course, but for these moments he was part of a proud parade of airplanes, each moving with a slow grace along the taxiway, through waves of shimmering heat created by the exhaust of powerful jet engines.

Many airline pilots acknowledge with a laugh that the hardest part of flying is getting away from the gate. Approaching the moment of takeoff, Captain Kevorkian was like so many pilots, eager to go. Finally, the time had come to fly.

At 8:18 P.M., Captain Kevorkian pushed the throttles slightly forward and the plane began to race down the runway. In about forty-five seconds the jet reached 184 miles per hour. Kevorkian pulled back on the control yoke, the W-shaped wheel mounted in front of him, rotating the nose of the plane up, and the 747 was airborne.

Moments later, as the landing gear receded into the aircraft, the plane was over Jamaica Bay. Turning east, Captain Kevorkian increased air speed to 287 miles per hour.

It was a busy time in the air around Kennedy. Departure control notified the TWA crew to make a second left turn to keep clear of another flight.

"The traffic in the turn will be three o'clock and five miles northeast bound four thousand feet," came a voice from the FAA tower.

Captain Kevorkian had been hoping to turn back to the most direct course, to "Bette," the first of a series of navigational way points the pilots would pass on their way to Paris, but the controller's instruction made it clear a turn like that would put the jumbo jet too close to a smaller Saab plane that was also ascending.

"He's at three o'clock?" Kevorkian questioned Snyder.

"Yeah," Captain Snyder replied, "that's the problem."

As the plane ascended, Susan Hill, forty-five, a police detective from Portland, Oregon, might have strained forward against her seat belt to see New York City out the window two seats away. Sailing above New York's overwhelming cityscape, Susan was introspective. Three months had passed since her divorce and she was embarking on a new adventure, her first trip alone. Hill had never needed a reason to be bold. She was impulsive and energetic. The tattoo of a ladybug on her wrist was evidence of her willingness to try something new, especially if she could shock the people who thought they knew her.

For her five-week trip overseas, she'd opted for a new look to go with her new life. Her blond hair cut short, she dressed in tight, sexy jeans and comfortable sneakers. The career girl clothes were stashed in the closet. Who knew if she would ever go back to that restrained professional look. Her backpack held a dozen new CDs from salsa to country and western.

Her plan was to house-sit at the Paris home of the brother of a fellow Portland police officer, then head over to England and spend a few more weeks there. Susan was well traveled. When she was married she'd visited India, the South Pacific, Singapore, and Russia. She'd never been to England or France.

This trip was Susan's second big adventure. Fifteen years earlier she had become a born-again Christian. She loved talking about it, so much so that she'd gotten herself in a little trouble. After she proselytized to suspects, her boss at the police bureau told her to keep her religious beliefs to herself.

This would not be a problem on Flight 800. Matthew Alexander, a twenty-year-old from Augusta, South Carolina, with a football player's build and a teddy bear personality, was seated next to Susan. He was en route to Dijon for a summer of Christian missionary work.

Seated directly behind them, fifteen-year-old Daniel Cremades was involved in his book. Cremades had spent three weeks in the United States, taking a college course for bright high school students at the Massachusetts Institute of Technology. Daniel was no egghead, though. He had an easy, self-confident manner that drew people to him, especially girls.

After his summer session at MIT, Daniel spent his final week in the states at the New Jersey home of his uncle and aunt, Dario and Jabina Cremades, and their eight-year-old son, David. Neither boy had a brother, so it was a thrill for both of them to be together. The two visited the museums in New York, just over the George Washington Bridge from the Cremadeses' high-rise apartment building, and spent their last day together at the Jersey shore.

Languages were so important to Daniel's parents, Jose Cremades and Ana Vila, Spanish citizens living in France, that every summer they sent Daniel and his older sister, Tania, to countries where they would be totally immersed in another tongue. The parents' efforts paid off. Daniel spoke four languages. For his flight back home, Daniel was carrying a book in English, Orson Scott Card's *The Abyss*.

Seated behind the wings were the honeymooners, Monica and Mirco. They'd been to New York, Los Angeles, San Francisco, and Santo Domingo in the Dominican Republic. They'd become enchanted with two aspects of American culture: Walt Disney and Hollywood. The day after arriving in New York, Monica called home to speak with her sister-in-law, Katia Buttaroni.

Last night we found a store filled with Disney animals, and I've already stuffed my bag with Mickey and Donald Duck and Dalmatians. And I bought something special for the baby, she told Katia, who was expecting her second child.

Six days into their honeymoon, Monica and Mirco called home again and learned that his grandmother had died. It was a devastating blow. Mirco was unusually close to his maternal grandmother. Guests at his wedding remembered that he had broken down in tears explaining that she was too sick to attend. The couple's first response to the news was to plan an immediate return to Italy. It took a lot of talking to convince them not to end their honeymoon then and there. Ultimately, they decided to continue on their trip, winding up on the beaches of Santo Domingo for their last week.

Monica went tropical and had her long hair cornrowed with tiny white beads woven into each of the hundreds of thin braids, an exotic hairdo for the folks back home. Monica enjoyed the quiet clacking of the beads each time she moved her head.

The last day of their trip, Monica called her mother. We're tired and ready to be home, she told her. We'll arrive in Rome tomorrow morning and take the train from there. I can't wait to see you.

As she sat by the window on the left side of the plane, Monica's hair was still braided, her wedding ring still shiny and new.

Flight attendant Ray Lang, fifty-one, had noticed the honeymooners when they boarded the 747. He was planning to be married himself, to fellow flight attendant Melinda Torche, who was also working the flight.

The night before, Ray and Melinda had dinner at a favorite neighborhood Italian restaurant in Long Island and made plans for what they'd do during their brief layover in Paris. After dinner, Ray held court at a spontaneous slumber party at the house he shared with his niece, Wendy, also a TWA flight attendant; his brother; and their mother. As they gathered in the room Wendy shared with Melinda when she was visiting, Ray predicted better times ahead for TWA and employees like them. I see things definitely lookin' up, he said.

The airline was emerging from its second bankruptcy in four years. The company had just announced a $25 million profit for the second quarter of 1996 and a longer term plan to update its fleet of aircraft, the oldest fleet owned by a major U.S. carrier, by replacing dated planes like the L-1011 and B-747 with the newer Boeing wide-body 767s. The airline was also adding twenty new long-range narrow-body 757s and fifteen McDonnell Douglas MD-83s.

The new jetliners were modern and sophisticated. The two-engine MD-83s, Boeing 767s and 757s consumed less fuel than the four-engine 747 and the three-engine L-1011, and a two-person cockpit crew saved the airline plenty in personnel costs.

Ray considered TWA's purchase a promising development. His faith in the reliability of Boeing was unshakable. "If it ain't Boeing, I ain't goin'" he'd told everyone gathered in the small bedroom. They all laughed in agreement.

When it came to enthusiasm for Boeing products, Lang had an ally in Captain Snyder, who loved the 747s he'd been flying for twenty-two years. In 1993 Snyder became a check captain on the wide-body, helping with the training of other 747 pilots.

On the night of July 17, 1996, Snyder was supervising Captain Kevorkian, a thirty-one-year veteran of TWA who had just become licensed by the FAA to fly the 747. As part of the airline's own training procedure, Kevorkian was completing twenty-five hours of actual flight time while the check captain, Snyder, reviewed and evaluated his performance.

Many pilots were intimidated by Snyder. His exhaustive interest in minimizing fuel consumption was just one example of his fastidious piloting. Snyder believed there was a good reason behind every routine flight procedure. Some pilots considered him a scold, while others attributed his "super pilot" persona to the breakup of his marriage in the seventies. Snyder's divorce was widely discussed because his wife, a TWA flight attendant, left him for another pilot. Since Snyder never remarried, his devotion to his job was noted and analyzed by everyone who knew his story.

As TWA 800 was pushing back from the gate, Snyder's attention to detail showed itself again.

"You got something else to do, Ralph," Snyder informed Kevorkian as the ground crew released the big jet's parking brakes.

"Number one ADP," Kevorkian replied, "and the electric," acknowledging that he knew the flight engineer needed to turn on the air-powered pump controlling the steering gear and ready the electric standby brake system.

Snyder wasn't content with the manner of Kevorkian's communication with the flight engineer.

"It's a command," he explained to Kevorkian. "That's a command," he repeated. It was a subtle adjustment, but Snyder's by-the-book correction left no room for a misinterpretation.

"Command," Kevorkian repeated. Changing his tone, he issued the order to the flight engineer, "Number one ADP on and the electric."

Before joining TWA in 1963, Snyder had spent four years in Cornell University's ROTC program and had flown for three years with the air force doing reconnaissance flights over Germany. His desire to establish a clear chain of command on the flight deck was less a function of his military background than of his often stated philosophy that good piloting required uncompromising clarity.

Kevorkian was not the sort of man who would have been stressed by a check ride with Snyder. The two had known each other for years, and Kevorkian knew what it was like to be in Snyder's seat as he'd supervised pilot performance himself when he was a captain on the wide-body L-1011, Lockheed's answer to the Boeing 747.

Sixty-three-year-old flight engineer Richard Campbell also had extensive experience with Snyder and 747s. He'd been a 747 captain until he turned sixty, the FAA's mandatory retirement age for pilots. After thirty years with TWA and a stint in the air force piloting the F-102 Interceptor, it was hard for Campbell to give up flying, so he began working as a flight engineer. At the airline, folks like Campbell are affectionately called "ropes," retired old pilot engineers.

Though Oliver Krick had been flying for TWA for only a few months, he was a member of the TWA family. His father, Ron

Krick, was a DC-9 pilot for the airline. Oliver's brother, Chris, had also expressed an interest in flying, and someday the three hoped to fly together.

Days before the flight, when Krick learned that he'd be on a trip with Snyder, he asked his father if he should be worried. The Kricks had gone golfing to celebrate Oliver's twenty-fifth birthday, and were waiting out a downpour in the car, trying to decide whether to go home.

How am I going to get along with him? Ollie asked.

Ron Krick chuckled at the question from his fun-loving son and recalled his own experience.

Look, I flew with Snyder on my first international flight to London. It was an eight-hour flight, but buddy, it felt like two. He was constantly asking me questions and making me look up the answers in the book. It wasn't the kind of stuff I learned in ground school, either, Krick continued, so my advice to you is, don't take offense. Listen up, and you'll learn a lot.

Ollie Krick depended on his dad's experience as a pilot from the time when, barely six years old and seated in his father's lap, he took the controls of the small prop plane Ron Krick used to give flying lessons. Nineteen years later as they sat in the steamy car waiting out a summer storm, Oliver Krick felt reassured by his father once again.

Had he known about that conversation, Captain Kevorkian would have felt some envy. When his son turned twelve, Ralph Kevorkian encouraged him to fly by paying for lessons. When Doug Kevorkian turned sixteen and his dad told him he would have to start paying his own way, Doug opted out and saved his money for a car.

Kevorkian was not surprised that his son had chosen wheels over wings, but he held on to his dream that his only child would one day follow him into the skies. Kevorkian could see Krick at the flight engineer's desk by simply turning his head to the right. Looking at Ollie, he might have wondered what it would be like if Doug were there instead.

Krick's job as flight engineer included making sure the fuel tanks empty uniformly. Feeding engines on one side of the plane with fuel drawn from the wing tanks on the other, an operation called cross-feeding, lets the engineer distribute the weight of the fuel evenly between the wings.

As Krick began this process with the right wing tank, he notified Kevorkian, "I'll leave that on for just a little bit." Then turning to Campbell to confirm his judgment, he asked, "Is that right?" To which Campbell replied a second later, "Yes."

Pumps in the wing tank began feeding the fuel through a large aluminum pipe running along the back wall of the plane's center tank. It surged through the line at a rate of approximately eleven gallons a minute for the five minutes and eight seconds that remained of Flight 800.

The cockpit activity was routine, almost mundane. In the passenger cabin, even nervous flyers would have been comforted by that.

The beginning of a transatlantic flight means at least three hours of nonstop work for the flight attendants before they can switch off the cabin lights, leave passengers to the movie, and sit down to their own meal. On this particular trip, they were looking at more than 200 travelers inconvenienced by an hour delay. As soon as the plane cleared 10,000 feet, Lang, Torche, and the others were on their feet, though the pilot had not turned off the "Fasten Seat Belt" sign.

From her seat by the forward galley, Judith Yee watched the cabin attendants beginning the process; filling ice buckets, preparing the beverage carts, and catching up on the latest gossip.

On the flight deck, the pilots increased air speed to 368 miles per hour. Krick radioed an "off report" to TWA's flight control office, giving the final and official push back and takeoff times and fuel load for Flight 800.

"Estimating Charles de Gaulle at zero six two eight," Krick reported. Six twenty-eight Greenwich Mean Time, early morning in Paris, around 2 A.M. by the pilots' body clocks. They'd all be ready for bed when they arrived.

Air traffic controllers notified the crew of a Beech 1900, a small regional airplane, flying south of them. The pilots searched the sky off the right side of the jumbo jet with a casual interest. The cockpit was quiet.

FAA rules forbid extraneous cockpit conversation from the time a plane pushes back from the terminal at the airport until it clears an altitude of 10,000 feet. The rule is intended to reduce distractions during a critical time. Flight crews have been known to engage in racy banter, but there was not a hint of questionable conversation on the flight deck of TWA 800. Between routine radio transmissions and discussions among the crew about the performance of their tasks, there were long silences as the men concentrated on their jobs.

The disaster was less than two minutes away when Kevorkian broke the silence. A quick spike on one of four engine gauges caught his attention.

"Look at that crazy fuel flow indicator there on number four," Kevorkian said out loud. He quickly scanned the other engine performance monitors for any indication that something was amiss, but could see none and sensed by the steadiness of the engines that fuel was flowing normally despite the reading on the gauge. Eight seconds later the vertical indicator tape made another hop.

"See that?"

There was no response from the other men on the flight deck.

The flight crew was probably unaware that the fuel flow gauge had been a recurring problem on N93119. There had been nine instances of incorrect readings during the previous two years. In some cases, trips to the hangar resulted in repairs; other times, no cause could be found for the faulty readings. Crash investigators would find this history significant, noting that intermittent electrical malfunctions can be symptomatic of serious wiring problems.

Pilots familiar with the aircraft and the crew flying it on this night believe that the event was considered nothing more than a minor in-flight glitch, one of many that could be expected on a machine as complicated as the Boeing 747. The jumbo jet is made of six million parts, and experienced captains don't spend too much time thinking about every widget that goes awry.

The crew was equally dismissive about trouble the refuelers had trying to fill the plane's wing tanks before departure. An automatic shutoff device designed to prevent overfilling had activated. Maintenance records showed eight reports of similar difficulty in the four previous months. On the afternoon of July 17, the fuel technician simply overrode the shutoff by pulling the circuit breaker to the volumetric control system and proceeded to pump tens of thousands of gallons of Jet A fuel into tanks on both wings.

When the plane reached its assigned altitude of 13,000 feet, Kevorkian leveled off and reduced power to the engines.

"Somewhere in here I better trim this thing," he muttered, mostly for the benefit of the pilot supervising his performance, and he started making the subtle adjustments to the rudder and wings to keep the 747 poised in flight in perfect balance. He'd just completed it when air traffic control gave clearance for another climb, this time to 15,000 feet.

"Climb thrust," Kevorkian directed Krick. There was no response. Seconds later he called again to the flight engineer to adjust the throttles to climb thrust.

"Ollie," Kevorkian called again, this time getting the young engineer's attention.

"Huh?"

"Climb thrust, climb to one five thousand," Kevorkian repeated.

Krick pivoted his seat frontward so that he could reach up to the console between the two pilots and pushed the four throttle levers forward.

"Power's set," he notified the pilots, maneuvering his seat back into position facing the engineer's panel on the right-hand side of the cockpit.

The whirring noise of the electric seat moving along its track was the last discernible sound picked up by the microphones in the cockpit.

4

Susan Hill, her seatmate Matthew Alexander, Daniel Cremades, Judith Yee, and Patricia Loo sat about one third of the way back in the passenger cabin, in seats bolted to a track on the roof of the plane's 12,890-gallon center fuel tank. Though they could not feel it, the living room–sized tank beneath their seats was hot.

The center fuel tank was designed as an extended range fuel tank, but it also absorbed the heat being generated by equipment below it, three enormous air handling units that turn blistering 450-degree air from the engines into climate-controlled air for the pressurized areas of the plane.

On July 17, these units were like stovetop burners under a saucepan, baking the remnants of the kerosene inside the tank into a crisp fog; an already volatile mix of fuel vapor and air was becoming more easily combustible with each degree increase in temperature. Two minutes after Krick turned the cross-feed switch, thirteen minutes after Kevorkian took the 747 into the skies, thirty seconds after increasing thrust for another ascent, this brew in the tank exploded.

The immediate damage was to the two forward-most walls of the fuel tank. In a flash, a crucial structural beam for the plane cracked in two under a force estimated at 100 tons of pressure. It took less than seven seconds for the plane's entire forward support structure to disintegrate.

All flight systems were severed and the cockpit decapitated from the rest of the plane. TWA Flight 800 fell into the Atlantic in a shower of light and a cloud of mystery.

Flying miles above the earth in unsurvivably thin air, air travelers take it for granted that they will breathe as normally on an airplane as they would on the ground. Pressurization technology that keeps the cabin environment oxygen-rich has been around since the late 1930s.

At 13,700 feet, the altitude at which TWA Flight 800 exploded, the outside air had half as much oxygen as the air inside the cabin. It would have been difficult for passengers to breathe. The quick drop in pressure would have caused pain in their ears. The noise of the destruction, strong enough to cause plates to rattle on shelves twelve miles away, could puncture eardrums and was very likely deafening. The outside air temperature, estimated at 35 degrees Fahrenheit, quickly chilled the cabin.

Those remaining conscious through the explosion would have experienced a numbing of all senses and a steep increase in heart rate from fear-induced adrenaline poisoning. That would be just the beginning of the hellish last seconds of their lives.

There was a racing wind as the thick cabin air rushed to escape through the ruptures in the fuselage. Until the explosion, air pressure inside the plane had been building with increasing force as the plane climbed, pressing up against every inch of the passenger cabin. Now it was unleashed like air bursting free from a punctured balloon.

Below the passengers' seats, tremendous pressure from the explosion in the tank lifted the floor in the area between the wings, slightly forward of the halfway point in the plane. Seconds later, a ten-foot hole appeared near the feet of

Deborah and Doug Dickey, chaperones for the high school students in the Montoursville French Club. Passengers in this area were sucked through an opening in the bottom of the plane. Dropping from two and a half miles in the sky, they smashed into the ocean at 120 miles per hour.

It wasn't just passengers falling. Overhead bins were popping open, contents ricocheting about. Even fixed interior components like the forward lavatory were ripped loose and pulled through the growing chasm in the floor. Chunks of the belly of the plane added to the heavy mist of debris.

At the same time, spidery fissures had raced up the walls at more than a mile a second, shredding ninety feet of fuselage into jigsaw puzzle–like pieces faster than the human eye could see it.

Monica Omiccioli was in a window seat, leaning against her husband when the tank exploded directly beneath them. In the subsequent breakup of the plane, this position somehow protected her from being torn apart and burned. Her new husband took the brunt of the assault. Susan Hill was torn to pieces. Daniel Cremades's face was damaged and his bones fractured. Only Judith Yee's bones were recovered, but her dog, Max, was not found. Mere fragments were all that remained of Matthew Alexander. Those who died from this severe and immediate trauma were the lucky ones.

With the area directly between the wings pulverized, the maelstrom was filled with pieces of the crumbling jumbo jet and the people onboard. High-velocity debris slammed into passengers. The medical examiner found pieces of the plane embedded in nearly all the victims. An upper-deck first-class passenger had been impaled through the abdomen by the components from an armrest. His body had been practically severed by it.

The destruction was so quick that much had happened before the event even registered with the cockpit crew. First there was the enormous blast of the center fuel tank rupturing toward them, followed so closely by the loss of electricity in the cockpit and the fog of decompression that it all seemed simultaneous.

Tearing metal; cables stretching, snapping, and whistling through the cabin; explosive thuds and shattering glass; screaming passengers—the dense cacophony was indecipherable. The crew was deafened by the noise and flying blind.

For moments, only a canopy of fuselage connected the cockpit and the first-class section to the rest of the jet. A giant crescent of lower fuselage was gone, leaving nothing to support the 18,000-pound nose. When it snapped off, it was propelled forward like a bullet until it lost momentum and began to arch down, picking up speed again as it headed toward the ocean.

Other airline pilots believe Captains Snyder and Kevorkian would have spent some of the last fifty seconds of their lives trying to fly the plane, instinct defying logic. Unaware of whether anyone could hear him, Snyder continued to talk into the microphone, documenting the chaos on the flight deck and his impressions of what was happening to the plane. Both Snyder and Kevorkian tried depressing the rudder pedals and pulling back on the control yokes in front of them, seeking a response from the plane, but it only continued its tumble through the air.

Braced in this way against the fall, four men with a cumulative century of piloting experience became passengers on their final flight.

The cockpit crew died of traumatic injuries when they hit the surface of the Atlantic. The pressure of impact on the right underside of the nose compressed it accordionlike all the way to the cockpit window line. Snyder, who very likely was not secured by all points of the shoulder, hip, and crotch belts of his safety harness, was blasted free of the plane.

Krick and Kevorkian went down with the ship, submerging slowly 120 feet to the smooth and sandy floor of the ocean. When they were found by Navy divers weeks later, Captain Kevorkian and Campbell were still strapped into their seats. Krick's brother Chris was told Oliver was intricately entangled in the aircraft wiring from the electronics bay under the floor of the flight deck.

"Embraced by his airplane," is how Krick's mother, Margret, thinks of it.

Once the front of the aircraft was gone, the center of gravity shifted, tilting the tail down. The engine fans, presumably still turning, drove the remainder of the 747 into a brief climb. This ascent was reversed when the outer ends of both wings snapped off.

As the plane began to fall, 24,000 gallons of fuel spun from the rupturing wing tanks, atomized, highly explosive kerosene clouds that quickly ignited, creating a huge fireball that was seen for miles. The right wing and a sizable piece of fuselage skin above it were burning when the wing folded up and snapped off. The left wing soon followed.

During all this, some passengers were shaken free of the plane entirely. Others were tossed out still strapped in seats that had been ripped off the mounting tracks. The pilots and passengers who were jettisoned with the front third of the aircraft had no way of knowing the plane had been severed, but the rear passenger cabin was still largely intact. Those passengers who were still conscious could see twilight where the front of the plane should have been and understood their situation with terrifying certainty.

These passengers were whipsawed back in their seats with the plane's ascent, then forced forward as it lost momentum and began to drop. All the while they were heaved from side to side with the uncoordinated tumbling of the jet in freefall.

The jumbo jet was generating enormous tongues of fire that were dropping from the sky along with eighteen four-foot tires and four two-ton engines. It continued to shed pieces and crumble from the gaping hole where the wing had been. A confetti of aluminum curlicues singed paths through the mild evening air.

The farther the tail section of the plane fell, the faster it went, so that when it hit the unforgiving surface of the Atlantic, it was moving about 400 miles per hour.

Through the narrow wraparound windows of the Eastwinds Airlines Boeing 737, Captain David McClaine watched the destruction of Flight 800 with his heart pounding.

Thirty seconds earlier, he estimated the plane headed directly toward him was a wide-body, probably a 747 or 767 starting a transatlantic flight. It was easy to see because its bright landing lights were still illuminated, even though the plane was probably above 10,000 feet, the altitude at which pilots usually turn out these lights. McClaine guessed the plane, ten miles in front of him, was TWA Flight 800, which he had just heard on the radio being cleared to ascend to 15,000 feet, still 1,000 feet below where he was flying.

Vinny, does that light look a little different to you, a little yellow? he asked his first officer.

Vincent Fuschetti looked up from the instrument panel and squinted into the sun, which was still shining off to his right.

Yeah, maybe. He shrugged. Fuschetti, who was piloting the plane, turned his attention back to adjusting the 737's course out of the path of Flight 800.

McClaine, however, continued to watch the plane, deciding, like a highway driver at night, to flash his own landing lights at Flight 800, as a reminder to the crew that their lights were still on.

His hand froze at the toggle switch. The 747 had exploded. He screamed at Fuschetti, My God, Vinny, what the hell was that?

The plane had become a blossom of flames, hovering briefly and then plunging, leaving dense black and orange trails above it. They watched the wings snap off, creating more flaming pillars as the wreckage spun toward the darkening sea.

McClaine's hand was trembling. He thumbed the button on the control yoke, cueing his radio mike to call to air traffic control using his airline designation Stinger Bee and his flight number 507.

"We just saw an explosion out here, Stinger Bee 507," the words tumbled out of him.

"Stinger Bee 507, I'm sorry I missed it . . . did you say something else?"

The voice of the controller sounded concerned.

"Ah, we just saw an explosion up ahead of us here, somewhere about sixteen thousand feet or something like that. It just went down"—McClaine paused—"into the water."

He could not take his eyes off the billowing trails of black smoke still suspended in the twilight. He was trying to understand what he'd seen. Are we seeing this? he asked Fuschetti. Then, Why are we seeing this?

There had to be three hundred people on that plane and we saw them die! McClaine continued, though his quick guess of the passenger load was significantly less than what a Europe-bound jumbo would normally be carrying in the heart of summer.

McClaine and Fuschetti could hear the controller at the long-range control center in Boston trying to raise TWA 800 on the radio. For nearly three minutes, the controller pleaded for a response. Finally, one came from McClaine.

"I think that was him."

"I think so," the controller agreed.

McClaine's voice had calmed somewhat.

"God bless him."

At least nineteen passengers survived the explosions and the torrential hail of objects loosened by the deterioration of the aircraft. Autopsy reports showed both smoke and water in the lungs of some people, indicating that they lived through the fall, breathing the smoke-filled air, and even briefly survived the plunge into the Atlantic.

For a long time after the last of the wreckage had settled, 800 pounds of glitter that had been in the plane's cargo hold rained softly over the crash site, leaving the debris and the floating victims coated in a glistening veil.

5

Hundreds of people in Long Island reported hearing the plane explode, but not one of dozens on the water in the vicinity reported hearing the impact.

Not twenty-year-old Chris Clapp, a college biology student with a deep tan who was sitting on his surfboard, not far from the beach at Fire Island, waiting for a wave and rubbing the stubble on his shaved head.

He was watching the sky, looking for the first stars to come out, when a speck of light he took to be an airplane caught his eye. It was pitching down and seemed to be performing an acrobatic dive.

Hey, check it . . . he started to say to the surfer paddling next to him, but the sentence trailed as the small white dot erupted. Instantly flames were falling in a wide, fiery curtain that began to separate into many burning columns.

He was still riveted to the brilliant display when he heard a deep rumble—the slow-moving sound of what he'd already seen. He felt a slight force push against his chest and the noise resonate in his rib cage.

Clapp remained in the water for twenty more minutes, watching as Coast Guard boats began to emerge from the inlet heading quickly east. Still straddling his surfboard, he was gently tossed in their wake.

As she hovered over the sofa, trying to get comfortable in her Westhampton beach home, the phone was wedged precariously between Naneen Levine's left ear and shoulder. When the pillows were arranged just so, she plopped into the nest she had made, facing a wall of sliding glass doors leading to her deck with a magnificent southern view of the Atlantic.

Levine was ready for a nice long chat, a conversation with her old college roommate, Kathy Tuchman.

Wow, someone's shot off a firecracker, Levine said, interrupting the story her friend was telling. There was a long pause. Kathy, I never saw a firecracker like that before. It's so small, just a pin of light, but it's really bright now.

There was another even longer pause followed by, "No, that's not fireworks. Lights are falling from the sky." Levine got up from the sofa and walked over to the glass doors. Tuchman could hear fear along with uncertainty in her friend's voice.

What is it, Nan? What do you think it is?

But the question just hung, unanswered. Levine seemed to be talking to herself, saying, This isn't good. I don't like this. Something's wrong, that's not fireworks.

Opening the glass door leading to the deck, Levine looked up into a cobalt blue sky turning orange-pink at the western edge with the fading sunset.

"Oh wow, that was strange."

Levine tried to redirect the conversation, asking her friend, What were you saying? and the two women resumed where they'd left off, but Kathy could tell Naneen's heart wasn't in the conversation anymore. Besides, Kathy's husband, CNN correspondent Gary Tuchman, had been paged and needed to call his office.

Gary's got to use the phone, Tuchman told her friend. We'll talk soon.

Levine said good-bye and went into the kitchen to hang up the phone. She was shivering as she walked back out onto the deck, checking the sky once again. But the only light she could see now was the final, feeble remains of the setting sun.

Long before the sound of the explosion had reached the ears of the people on Long Island's south shore, the ignition of the 747's fuel tank turned the airplane into a bright white beacon high in the sky. It took the boom fifty seconds to reach land, two miles below and nine miles to the north of the plane's final course.

Members of the 106th Rescue Wing of the New York Air National Guard at the Francis S. Gabreski Airport were alarmed by a flash of light and a fire in the southern sky, but described the silence that accompanied it as surreal.

In a grassy area near the airfield, parachuters Shaun Brady and Craig "Jake" Johnson saw a giant fireball light up the sky. They were fearful they'd witnessed the crash of the C-130 transport plane from which Johnson had just made a practice jump.

Brady keyed his radio microphone and called to the pilot. Reading Brady's mind, Major Mike Weiss quickly responded, "Negative. It's not us. We see the fire as well . . . we're heading to find out."

Two hundred feet above the Gabreski Airport runway, National Guard Captain Christian Baur was flying a Sikorsky Pave Hawk helicopter when he spotted a quick white flash in the sky to his left.

Denny, does that look like pyro to you? the young pilot called back to his friend, fifty-year-old flight engineer Dennis Richardson.

Baur was confused. His first thought was that Weiss's C-130 was tossing flares as part of their practice search and rescue operation. The thought barely had time to register. Richardson hadn't answered the question when the small light became a huge horizontal fire zipping from right to left across the sky.

Oh my God, do you see this thing? Baur cried, continuing to look at the sky where the pin of light had been just a second earlier.

What the hell is that? Richardson hollered, leaning forward between Baur and Major Frederick "Fritz" Meyer, the pilot in the right seat. All three men were transfixed by the enormous line of fire that was beginning to drop like a curtain falling until a wall of flames was practically all they could see out the wide front window of the Pave Hawk.

Meyer, a highly decorated Vietnam veteran with a handsome chiseled face and ice blue eyes, was monitoring the radio transmissions between the unit's aircraft and the Gabreski air control tower. He could hear Weiss, on the C-130, remarking on the same horrific scene.

"Notify the Coast Guard. It looks like two planes have collided. We're heading over," Weiss told the tower.

Major Meyer had not said anything to Baur or Richardson about the explosion or what they'd witnessed together. So the men were shocked when he mumbled into the headset the crew uses to communicate with one another through the din of the noisy aircraft, "We're outta this. We gotta go back."

Whadda ya mean "back"? Baur snapped.

They want us to go back, was Meyer's cryptic response.

Who's they? he asked. Baur was incredulous. Meyer was the commanding officer on the flight, but Baur, a scrappy, irreverent New Yorker with more confidence than patience, was at the controls. He maintained a course to the crash scene, cutting off Meyer by insisting, We have to check it out.

As they approached the plume of smoke, Richardson was startled to hear Baur shout, Whoa! and quickly maneuver the helicopter into a tight right turn away from the site.

What's up? Richardson asked. There was no need for a reply. Through the window it looked to the flight engineer as if someone had suddenly dumped a giant trash can over the aircraft. Embers, paper, plastic, and foam billowed gently down from the sky.

"Stay back, there's still a lot of falling debris," Baur was warning the C-130 over the radio.

What Baur did not say out loud was that among the hail of unidentifiable chunks he'd seen a person. "Falling right through the debris," he recalled later, "like a sack of potatoes."

The pilot of the C-130 heeded Baur's warning. Through the many windows on the giant transport's flight deck the pilot, Mike Weiss, could clearly see a column of smoke streaming up from the surface of the Atlantic. It was so dense he wondered if the explosion had occurred on the water. Perhaps a tanker had blown, and the fire was erupting upward like a volcano.

Not long after, the men could identify pieces of an airplane tail section, one white, one red, settling in the water. A massive oil slick about the size of two football fields was on fire. As more fuel percolated up, larger areas on the surface of the water were ignited.

There were six men on the transport plane, about 1,500 feet above the scene. All eyes were searching for a glimpse of survivors in the flaming sea.

Weiss saw two semi-inflated evacuation slides, creating an oval shaped area of disturbed water. He recognized the chutes right away.

"This has to be a passenger jetliner," Weiss radioed to the control tower. "Call the authorities."

"Do you think there are survivors?" came the reply from the tower.

"No," Weiss answered, and to himself he murmured, "Not a chance."

Minutes later Weiss's bleak prediction was confirmed. He spotted an empty life raft and twenty to thirty bodies floating together in a clump nearby.

. . .

Sixty-five miles away, FBI deputy director Jim Kallstrom stood on the sidewalk with a small group of fellow cops, taking advantage of the mild summer evening. The men had just left a dinner honoring Raymond Kelly, New York City's former police commissioner, who had joined the Treasury Department. Kallstrom, a twenty-eight-year crime fighter, rarely got a chance to let down his guard. In the presence of other law enforcement officers, he relaxed and lingered in the twilight shadows of midtown Manhattan's skyscrapers even after the party was over.

Glancing at his watch, Kallstrom remembered he'd told his wife he'd try to make it an early night. He said good night, got into his car, and called her with the news that he was about an hour away from home. Before he'd ended the conversation, his pager went off. Still parked at the curb, dialing the FBI night duty officer, he heard more beepers going off on the hips of the men remaining on the sidewalk.

At home in Connecticut, Susan Kallstrom's phone rang again the instant she ended the conversation with her husband.

Sue, it's Charlie. It was Charles Christopher, an FBI agent in the New York bureau. Charlie and his wife, TWA flight attendant Janet Christopher, were longtime family friends of the Kallstroms. He was sobbing as he said to Susan, Janet's plane has gone down. It's crashed. I need Jim. We've got to look for her.

Coast Guard Seaman Ken Seebeck was thinking with his stomach. It's eight thirty-five. If we hurry we can get back to the station before mess ends, he told fellow Coast Guardsman Jarl Pellinen. Having spent the past several hours towing a sailboat from the dead calm waters to the Shinnecock station, Seebeck and Pellinen were eager to get back to their own station at East Moriches, about eight miles farther south down the bay.

Seebeck looked over at the stationhouse to wave good-bye to petty officer Rick Freese, who was at the window. Ken, there's something going on at Moriches. You better get back there quick! Freese hollered at Seebeck.

Seebeck and Pellinen stopped and turned toward Freese. What've you got? Seebeck asked.

Don't know, just go!

The two men trotted down to their twenty-eight-foot Monarch utility boat tied up at the dock; Craig Achilli and Sam Miller were already aboard. Pellinen shouted to them as he approached, Gotta get back to East Moriches, something's up.

Over the marine radio distress channel, the seamen heard the order at the same time. The men prepared the boat quickly and revved the twin outboard engines. When Seebeck and Pellinen jumped aboard, they pulled away from the Coast Guard station. Without even realizing it, they'd become energized by the urgent tone of the boaters' emergency calls reporting explosions, distress flares, falling stars, and columns of smoke not too far from their own Coast Guard station.

For a while, the men heard only the radio and the boat's engines. Forty-five minutes later, they heard the distant sound of helicopter rotors and saw flares dropping from the sky. Seebeck grabbed the radio microphone and called back to Freese at the Coast Guard station.

There's an Air Guard plane dropping flares, he reported.

Switch to secure frequency, Seebeck, responded Dan Phee, the Coast Guard officer responsible for all seven Coast Guard stations along the Long Island coast. Seebeck adjusted the radio frequency to the correct channel. Phee gave him the coordinates, telling him to race to the accident site.

The Air National Guard is on the scene with what appears to be a commercial airliner in the water, Phee said in a controlled voice. The cutter *Adak* is there and will be in command, he continued. Phee's voice changed in intensity and became softer as he told Seebeck, There are numerous people in the water. Prepare your crew.

Roger, Seebeck replied. He placed the microphone back on its hook, wondering what a commercial plane crash would be like. None of the men on the small boat, including Seebeck, had seen that kind of disaster before. He was as unprepared as his crew.

Seebeck leaned his head out of the window of the pilothouse. It's a commercial airplane crash. Let's get the EMT kits out, he shouted. We'll need gloves, protective equipment, oxygen. Get the spotlights, too. After assembling the gear, the men took lookout positions on the boat. Off in the distance a blurry orange glow appeared on the water.

A piece of wreckage, identifiable as part of a commercial jet by its size and paint job, jutted from the water. A wall of fire leaped thirty feet in the air, catching Seebeck off-guard. Oh my God, he murmured.

There were few men in Washington with whom National Transportation Safety Board chairman Jim Hall felt as comfortable as Peter Goelz. Hall, a slight, silver-haired Tennessean with a slow drawl and a quick temper, was always impeccably groomed. Goelz, his New York–born and bred public information officer, struggled just to keep his shirttail tucked in. Nevertheless, the two men shared a goal: to take Hall from Tennessee politics, where he'd once worked for Governor Ned McWerthor and Albert Gore Sr., to the national stage. With careful maneuvering, his appointment to head the once obscure safety board could be used to cultivate a reputation as a champion of safety. One and a half million people step onto U.S. airliners every day, so the constituency was obvious.

On the night of the crash of TWA Flight 800, Goelz and Hall were with Ken Jordan, the NTSB's then-managing director and Hall's longtime friend, and their wives at a movie. When the pagers of all three men went off simultaneously, it was Goelz who got up to find a phone; Jordan and Hall soon followed.

In the lobby, Goelz hung up the phone and announced, There's a 747 down off the coast of Long Island. Two hundred plus aboard. There's a search and rescue going on. It doesn't look good.

The men headed back to the safety board headquarters not far away. The short taxi ride was silent, each man pondering what would need to be done to launch an investigative team to New York.

From Ken Seebeck in the coast guard Monarch came the call, This is the UTL280-501 to *Adak*, we are on scene and requesting tasking. The cutter *Adak* was in command, Seebeck and his men were waiting to be told what to do. The water was illuminated by the light of the burning plane and their own high-power, hand-held searchlights. The men had been looking for survivors. All they'd seen were aircraft debris, bits of human organs, and charred body parts floating by in a fog of smoke.

UTL280-501, this is the *Adak*. Get in as close as you can to the wreckage without jeopardizing your vessel and search for survivors.

Nosing the craft into a hollow between a piece of tail sticking high out of the water and the large fuselage they'd seen initially, Seebeck began making electronic entries on the boat's navigational system at every point where the crew saw human remains. This would make it easier to find them when it came time to recover the dead. For now, they were focused on finding survivors.

The men were so absorbed in this job that they didn't anticipate the danger of running the motors through the thick stew of airplane wreckage until it was too late. Suddenly the engines stopped and the boat stalled.

We've fouled the screws! Seebeck yelled back to the crew. Let's clear them out. He pulled on the lever to raise the twin outboards above the water. From the pilothouse, he notified the *Adak* that the outboards' propellers had gotten tangled and the Monarch was adrift. He radioed his location to the *Adak*, hoping the cutter could make its way over and tow them out before their boat drifted into the fire.

It was quiet on the water. On the tailfin of the Monarch, the men cursed as they tried to get a grip on the oily mess that had twisted around the propellers. Freed wires sprung loose, shooting like projectiles across the deck.

The men tried not to focus on the raging fire nearby, but the heat on their skin told them they were close. Every breeze brought with it the stench of burning jet fuel and the sound of crackling flames.

From the pilothouse, Seebeck asked the *Adak* for its position. He was trying to remain calm, but his boat was only feet from the fire. His men were already coughing from the smoke. Achilli's shout from the back of the boat drowned out the radio response from the *Adak*.

We're clean, Ken, give it a try. Hurry!

Seebeck lowered the engines and turned the ignition key. The Monarch jolted to life. Seebeck threw the boat in reverse and roared away from the fire.

After seeing the explosion of TWA Flight 800 from the drop zone at Gabreski field, parachuters Sean Brady and Jake Johnson raced back to the 106th Air Rescue Wing building and hastily began gathering emergency medical supplies, wet suits, survival vests, and night-vision goggles. The Pave Hawk helicopter was back to pick up the parajumpers. Since the last daylight was gone, Major Mike Noyes, a helicopter pilot certified to fly night missions, took over for Meyer.

As the helicopter lifted off, Brady and Johnson donned headsets, stowed the emergency equipment, and carved out space for survivors along the cramped area at the back of the aircraft.

Brady could hear the pilot telling air traffic controllers that he had the wreckage in view.

"It was huge, bigger than anything you can imagine," Brady recalls of his first sighting of the crash scene.

The Pave Hawk was flying perilously low, at times just fifty feet off the surface of the water, to offer the rescuers the best view. Wearing a full-body wet suit and ready to dive if necessary, Brady lay belly-down on the floor, extending his face over the edge of the doorway outside the helicopter.

The 106th Air Rescue Wing's primary responsibility is search and rescue in combat situations. Pilots and parajumpers are trained to work in low-light operations, flying in darkened aircraft, using night-vision goggles that multiply available illumination by a factor of thousands. When Coast Guard and other rescuers arrived, their blazing searchlights threatened to blind the National Guard crew. With its lights blacked out, the Pave Hawk was virtually invisible to the arriving aircraft.

Noyes carefully piloted the helicopter to an area away from the spotlights and flares, avoiding the flames and protruding wreckage. Baur repeatedly updated the helicopter's position on the radio for every other pilot to hear and heed.

Oblivious to the worries of the pilots, pararescuer Shaun Brady cupped his hands around his mouth screaming through the racket of the shuddering helicopter as body after body submerged lifeless under the sea spray whipped up by the rotors.

"Somebody be out there! Just one person be out there!" He was unnerved by how unharmed many of the victims appeared, as though they'd just fallen asleep in the waves, yet others could barely be identified as human.

In 1991, Brady, a former marine working in his parents' Long Island restaurant, was inspired by the story of Rick Smith, a pararescuer who had died a hero when his Air National Guard helicopter had ditched in the stormy seas after trying to save a boat during a storm in the Atlantic. The story was dramatized a few years later in Sebastian Junger's bestseller *The Perfect Storm*.

Brady had marched right down and enrolled in the Air Force Special Forces Training Academy, spending 1992 in rigorous training and the years that followed on various rescue missions around the world. The crash of TWA Flight 800 was the worst disaster he'd seen.

Unable to find a single survivor, he tossed red and green chemical light sticks out of the helicopter to mark the location of the dead for the boaters to retrieve. Soon there was a dense starscape on the surface of the water, testament to what he couldn't accomplish.

Returning to base three hours later, he went to a nearby bar and promptly got drunk.

6

Ray Lang, he's dead.

Hearing no response, Michael Kempen, a TWA flight attendant, turned from his computer screen, which displayed company scheduling information, to look over at his girlfriend and fellow flight attendant, Laura Beth Miller. The two of them were in the loft of the apartment they shared in Queens, not far from Kennedy Airport. She, too, sat facing a computer screen. The sound of the news announcers on television had obliterated the sound of her sobs, but Kempen could see her shoulders heaving, and he knew she was crying.

He had the strange feeling that he was simultaneously living the moment and observing it from a distance, a sensation that continued even as Miller, her face red and covered with tears, turned toward him and began reciting names. Jim Hull, Olivia Simmons, Barbara and Lonnie, Elias, Daryl ... My God, Michael, everybody's on that flight!

Their eyes met and they were silent for a moment, a fleeting communion soon broken by the ring of the telephone. Their friends were calling to find out if either Michael or Laura, who'd worked for the airline for twenty years, had been on Flight 800.

Miller picked up the phone. Kempen turned back to the monitor to continue checking the crew lists and work schedules. He'd had access to all that information just minutes earlier, but now it was gone.

"What's happening to the files?" he wondered to himself.

At least an hour passed from the time Kempen heard about the crash until he remembered he was part of the TWA crisis response team, asked to volunteer because of his work in the past assisting in his union's drug and alcohol abuse program. At 10:30 P.M., Kempen went to Kennedy Airport to see if there was anything he could do to help.

When he arrived at Hangar 12 fifteen minutes later, no one at any of the TWA offices knew what the crash response plan was. Kempen returned home shortly before midnight.

Bob Golden had just returned to his Long Island home from a New York Yankees game. He was headed for his bedroom, pulling off his shirt, when the phone rang. It was Bob Boergesson, the civilian dispatcher on duty at the Suffolk County Medical Examiner's Office. His wife, Kathy, followed him into the bedroom, snapping on the television as Bob talked on the phone. Kathy usually kept one ear on her husband's phone conversations. Fourteen years of marriage to an investigator in the medical examiner's office made her alert to calls that came in the night. This time she was more intrigued by the TV news bulletin. She barely heard her husband saying, East Moriches? I'll get right over. She saw him grab a clean shirt from his bureau and quickly put it on. Only then did she realize the horrible news from the television set was drawing her husband out into the night.

It took Golden less than five minutes to get from his house in the Long Island southern coast community of Center Moriches to the East Moriches Coast Guard Station, located on the far edge of a grassy spit of land. The brief trip had unnerved him. Dozens of emergency vehicles raced past, lights bouncing off the small neighborhood businesses locked up for the night. The blaring of sirens and whirring of helicopter blades was relentless. When he arrived at the field in front of the Coast Guard station, he smelled boat fuel mixed with the unique aroma the marsh gives off on summer nights. The smell would only get worse.

It was 3:15 A.M. in Paris when the ringing sounded loud and urgent. Patrice Soualle sleepily reached over to the nightstand and patted the snooze button. When the noise continued, he realized it was the telephone, not the clock.

The call was shocking but not surprising. Since becoming TWA's station manager at Charles de Gaulle Airport ten years ago, Soualle had wondered "what if."

"Every station manager has a little black cloud over his head. Inside you're saying, 'Am I ready to face an accident?'" Fully awake within seconds of hearing that TWA Flight 800 had crashed with no survivors, Soualle dressed and prepared to leave for the airport, a fifteen-minute drive from his home in the Paris suburb of St. Witz.

Before leaving, he shook his still sleeping wife and put her to work; Lydia, here's a list of people you must call for me. Please let them know we have an accident and they should come to the airport. Call everyone you can call."

Lydia reached over to turn on the light by her side of the bed, noticed the time, and then kissed her husband good-bye.

Soualle intended to get right to work when he got to his office. He knew just what to do. TWA and three other American carriers with operations at Charles de Gaulle Airport had created a plan for aviation emergencies, following the bombing of Pan Am 103 over Scotland.

At his desk, he found it hard to direct his thoughts. In his mind, he could see the giant red-and-white 747 slowly sinking into the water. He commanded himself to pay attention to the enormous list of jobs to be done, not the least of which was handling the families. Dozens of people would soon be arriving at the airport to greet their loved ones.

"Focus on the mission. Handle the families, attend to their needs," he told himself. "There's nothing else you can do."

Bob Golden's first job at the East Moriches Coast Guard Station was to prepare a makeshift morgue and design a system so that a hastily assembled staff could process several victims at the same time. At midnight, he received word that the boats were approaching the station with bodies.

Golden stood on a chair to brief those present on what they were about to do. As the bodies arrived, each was to be tagged with a number and photographed. Personal effects were to be listed on a slip and placed along with the corpse into a zippered body bag. The second tag with the same number would be placed on the outside of the bag and the zipper locked.

He created four teams of four: a forensic scientist or physician's assistant from the coroner's office to examine the body, a homicide detective from the Suffolk County Police Department to take notes, an ID detective to photograph the body, and an ambulance driver to assist as needed.

His staff was used to handling victims of car accidents, drownings, fires, even murders, but few had experienced the carnage of a commercial airline crash.

"I deal with bodies one at a time, maybe two," Golden said later. "Never so many, so much death on this scale. For a while I worried that bodies would be arriving faster than I could handle them."

Fortunately, Suffolk County had a disaster plan. It had not escaped the attention of the county's new medical examiner that the area was in Kennedy's flight departure path.

Dr. Charles Wetli had come to Long Island in February 1995 after fifteen years as chief medical examiner in Dade County, Florida. He'd been through the destruction of Hurricane

Andrew, which was blamed for more than 100 deaths. His experience gave him a perspective Suffolk County's other emergency agencies hadn't fully considered.

"Everybody thinks of mass disaster as saving lives, but what happens if everybody dies?" Wetli asked.

Dr. Wetli has the bedside manner of a doctor whose patient is always dead. His complexion is pale, his manner stiff. He has a slightly uncomfortable physical presence that is compounded by his tendency to speak bluntly. His colleagues admire him; strangers find him arrogant.

Countering the hostile fire directed at Dr. Wetli would become Bob Golden's next and even more difficult job, and he would not be successful.

On the night of the crash, the medical examiner's teams, sweltering in biohazard suits and surgical masks, processed the bodies in a grim assembly line. Seven hours later, 100 victims had been tagged and taken by refrigerated trucks to the morgue forty miles away in Hauppauge, New York.

On his second trip carrying supplies from the car to his boat, WNBC television news reporter John Miller was startled and then delighted to run into Tony Villareale at the otherwise deserted marina in Westhampton.

Tony, you're just the man I'm looking for! he exclaimed, shifting a load of jackets and blankets to his other arm and steering the balding former wrestler through the cool, dark night toward the twenty-four-foot Boston Whaler tied up in its slip. The previous fall, Villareale had talked Miller into purchasing the fishing boat; now Miller needed Villareale to sail it.

You sold me this boat so you know I'm not even that good in the daytime, he said. Help me get out to the scene of that crash.

You serious? Villareale replied. Not waiting for an answer as he stood shivering on the dock, he said, You got a jacket or something?

Yep, Miller said, pulling a jacket from the bundle he'd just brought on board and tossing it up to Villareale.

Okay, let's go, Villareale replied, pulling the windbreaker on over his T-shirt and shorts and jumping into the boat.

Steering the Whaler down the shallow two-mile inlet leading to the Atlantic, Villareale had no idea how he'd find the crash scene. But Miller exuded a big-brother confidence that Villareale found reassuring. From the captain's seat he turned around to see Miller opening a Halliburton case containing an astounding array of electronic equipment: a cell phone, video camera, a two-way radio, ship-to-shore radio, police scanner, batteries, battery chargers, and cigarette lighter adapters.

Over the marine radio, Miller and Villareale heard a dispatcher requesting all small craft to respond to the crash scene and render aid to victims. Both silently wondered if they'd have the chance to be heroes. Speeding along Long Island's ragged coastline, Miller and Villareale were confounded by how dark and peaceful the night was in contrast to the frenzied communications on the marine radio.

"I didn't know if we were gonna find it," Villareale remembers. "I didn't want to find it, but John was determined. He was the reporter. I'd sold him all this fancy equipment for the boat like GPS (global positioning system), so I told him to call his station and get the coordinates, which he did."

Villareale turned the boat south, deeper into the open water of the Atlantic, where everything changed twenty minutes later. The water remained still, but huge tongues of flame were leaping from the surface.

Above, helicopter rotors were making a syncopated growl, backed up by the bone-penetrating rumble of the giant Air Guard transport plane flying low passes. The outboard engines of a dozen idling vessels, including pleasure boats, added to the din, along with the shouts of would-be rescuers who were calling out for assistance with the corpses.

It was a gruesome task, emotionally and physically. The victims' bodies were water-soaked, leaden, and slippery. And they were fragile. At times, limbs would separate from torsos, coming off in the hands of the boaters who had, without warning, become part of this living nightmare.

Occasionally one of them would shout obscenities at Miller, who had pulled out the video camera and was now recording everything.

For much of the night, Jim Kallstrom was on the telephone. He'd arrived at FBI headquarters in lower Manhattan shortly before 9 P.M., still shaken by the news that Janet Christopher had been on Flight 800. Kallstrom had been part of the wedding party when Janet and Charlie married. His sense of sorrow and dread mixed with the adrenaline rush he always felt when tackling a major crime. He was already certain this crash was a criminal act. After briefing his boss, FBI director Louis Freeh, he began preparing a plan for dealing with the press.

Kallstrom prides himself on his ability to meet reporters' needs while actually satisfying his own. By letting the press in on his suspicions, he could kick-start what was clearly going to be a massive investigation.

We want people with any information to know where to call, he told his public information director, Joseph Valiquette. Get the phone company to get us a toll-free number right now. Then get it out.

With the conspiracy trial of Ramzi Yousef going on in the Federal courthouse just across the street from Kallstrom's office in lower Manhattan, law enforcement quickly presumed the TWA disaster was the work of terrorists.

"We made that assumption based on what we knew," Kallstrom said later. "First, the plane blew up. We had almost instantaneous reports from other airplane crews. There was no communication from the cockpit, there was no distress call. And this business of people reporting that they'd seen things streaking in the sky, just those facts alone were enough."

By the age of fifty-three, Kallstrom had a well-developed confidence in his own opinions, and his authoritative manner swayed plenty of others, if not to his way of thinking, at least to reconsider their own.

With the wreckage of TWA Flight 800 still burning in the Atlantic, the FBI's Kallstrom knew he was looking at an act of war. His gut feeling was reinforced by his calls to the National Transportation Safety Board in Washington, the federal agency charged with the investigation of airplane crashes. No one at the NTSB could remember a case in which a plane had simply exploded without warning from a mechanical malfunction.

Kallstrom promptly announced to reporters that the FBI would be investigating the incident as a potential criminal destruction of an aircraft, a federal crime. He immediately launched agents to look into the possibility that a missile took down Flight 800. Tom Pickard, Kallstrom's deputy for national security matters, was assigned to find out what kind of missile could destroy a 747.

Through the night, Pickard pressed defense, aerospace, and weapons specialists for information about whether a missile exists with the capability of hitting a target two and a half miles up. Certain types could, he learned, but probably only when fired from a boat on the water, not from land, nearly ten miles to the north of the flight path.

The information was intended to narrow the investigation, but instead it opened it up. The operators of hundreds of vessels, commercial boats, barges, and pleasure boats were now under suspicion. If a missile had been fired from the water, the suspect would already have a night's head start.

The FBI also considered the possibility that the plane had been hit by friendly fire. The area directly to the south of TWA's flight path was an active weapons fire zone, a 300-mile-wide swath of airspace regularly reserved by the military. The zone, called Whiskey 105E, was for airspace below 4,000 feet, more than a mile and a half below where the 747 was flying. On July

17 there were extensive military maneuvers under way in Whiskey 105E involving Navy ships, submarines, and aircraft.

Call the Pentagon and find out what they were up to tonight, what military assets were in firing range, Kallstrom ordered Pickard. Don't they have some air corridors reserved for firing weapons? Who's in charge of that anyway, the FAA or the Pentagon? Find out.

Kallstrom, a former Marine, thought the possibility of the U.S. military being responsible for the crash was remote. Nevertheless, he considered it worth a look.

What do you mean there are no pilots? Peter Goelz's voice boomed across the room, causing the small crowd assembled in NTSB chairman Hall's office to stare at the usually placid public relations man.

It was midnight, and safety board investigators were eager to get to what appeared to be one of the largest commercial air crashes in the nation's history. Though the Federal Aviation Administration in Washington, D.C., keeps a sixteen-passenger Gulfstream jet on standby for just this reason, it had failed to schedule an overnight flight crew. Pilot sleep/rest restrictions are intended to enhance safety. Now they were keeping the investigators on the ground. The irony of the situation was not lost on anyone in the chairman's crowded office.

What's the point of having a plane on standby for the NTSB if there's no one standing by to fly the goddamned thing? Goelz bellowed.

By now, the board's vice chairman, Robert Francis, had arrived. Francis, a balding man with bright blue eyes and a quiet, almost sleepy manner, was the board member on call at the time of the crash. It would be his job to accompany investigators to the scene and act as spokesman for the NTSB. TWA Flight 800 would be Francis's second major airline crash in just three months. He'd been on the scene in Miami following the

harrowing in-flight fire and crash of a ValuJet DC-9 in May
that had killed all 110 aboard.

Francis's assistant, Denise Daniels, and NTSB deputy director of the office of aviation safety Ron Schleede, were working
the phones along with Goelz, lining up rental cars, hotel
rooms, and cellular phones for the team and passing on information to Alfred Dickinson, who would be the investigator in
charge.

Several television sets had been brought into Hall's two-
room office suite, and TV news announcers filled every silent
pause. Aviation experts were presenting a jumble of theories
about the crash in the absence of official information, and
"unnamed federal law enforcement sources" were reported to
be looking at the crash as a likely act of terrorism.

It was clear to everyone in the office that the NTSB, charged
with determining the cause of the accident, was already losing
control of the investigation under an avalanche of speculation.
Hundreds of law enforcement officials had been on the scene
for hours. The official investigators hadn't even found a way to
get there.

The television news that night did not help to explain what
brought down the jumbo jet. To an airplane wiring specialist
named Ed Block, watching TV in a Philadelphia suburb, the
news had an eerie familiarity.

In ten years as a aircraft wiring expert for the Pentagon,
Block, forty-six, learned about military jets blowing up. It was
no mystery to armed forces investigators what caused many
fighter planes to crash. Certain types of insulated electrical
wire failed to isolate the current, and resulting short circuits
had caused fires and, sometimes, catastrophic explosions.

In 1981, when it became clear that substandard wiring
insulation was responsible for a rash of Navy fighter jet accidents, Block was knowledgable but inexperienced. He was

learning that a startling number of F-14 accidents, 20 percent of 600 planes, had developed some sort of electrical trouble, turned Block into a zealot. He directed all his energy into ridding military jet fleets of the suspect wiring and warning manufacturers Boeing and McDonnell Douglas not to use it on commercial planes.

On the night of July 17, 1996, Block prayed for forgiveness for his first reaction to the crash of TWA Flight 800, which was a feeling of vindication. He was certain the disaster was what he'd been warning about all along.

As dawn broke over Long Island on July 18, Kallstrom and Pickard sat in a National Guard Pave Hawk helicopter, overwhelmed by the events of the previous ten hours. A night of analyzing the information available had reinforced Kallstrom's earlier conviction that the plane had been the victim of terrorists. But who? And how? And was TWA 800 just the first strike? The thought of additional attacks was chilling.

The Pave Hawk slowed and hovered about 500 feet above the ocean. An acrid, burning smell reached the helicopter where men were surveying a field of aircraft debris and personal effects bobbing on the water. Kallstrom looked around in all directions, but the littered surface extended to the horizon.

After a minute Kallstrom turned to Pickard, with whom he had worked for over twenty years. What are we gonna do? Kallstrom asked, ratcheting up his voice over the din.

Taking the question literally, Pickard shouted back, We've got to collect every piece of this, thinking of the crash of Pan Am 103 and how a small piece of wreckage provided the key to what happened.

Kallstrom turned back to the window. After five minutes, he directed the pilots to take him to the Coast Guard station at East Moriches. He had an appointment at 7:30 A.M. with the NTSB's Francis and did not want to be late.

7

At dawn, Al Dickinson, the NTSB's investigator-in-charge, and a dozen others finally boarded the FAA jet at Washington's Ronald Reagan National Airport. Few had gotten more than a brief and restless night's sleep.

On the plane was Deepak Joshi, forty-two, a small man with a dark mustache, jet black hair, and the conservative way of dressing that typifies many engineers. He was eager to please and able to take charge. His attention to detail was helped by a near compulsion for organization. The latter qualities would turn out to be particularly important in this investigation, where pieces of wreckage would number in the millions and the documentation would weigh tons.

Assisting Joshi was Frank Hilldrup, a young engineer with Tom Selleck looks. Hilldrup was a hero within the NTSB for supervising the deep underwater recovery of the black boxes from a Boeing 757 charter that crashed into the Caribbean shortly after takeoff from the Dominican Republic just six months earlier.

Cockpit voice recorders (CVRs) and flight data recorders (FDRs) are housed in nearly indestructible metal boxes that, contrary to their common name, black boxes, are painted bright orange for visibility. These recorders are positioned in the tail section in most aircraft. Information about the plane's operation and systems during flight and the recording of the last thirty minutes of cockpit conversations have often helped determine what caused a crash. Wreckage is revealing, but in a fatal crash, recorders are the posthumous voices of planes and pilots.

Recovering black boxes is nearly always the highest priority in a crash investigation. It was no different in the crash of TWA Flight 800. Despite the difficult start, experienced NTSB investigators, "tin-kickers," as they're called, were optimistic that once the black boxes were found, the mystery would be solved. It was days before they realized the only luck they'd have on this investigation would be bad.

First, the recorders were agonizingly elusive. When they were finally recovered one week later, the information was not illuminating.

Then there was the crash scene itself. The plane had been shredded and strewn over a seventy-square-mile area, then submerged under 120 feet of water, where it concealed the remains of more than 100 passengers.

The first hint of trouble came even before the NTSB's tin-kickers left Long Island's MacArthur Airport.

It was expected that NTSB vice chairman Francis would travel to the scene with the investigators, but he left ahead of them. Dickinson didn't see him again until he arrived at the Coast Guard station in East Moriches and found Francis involved in a planning meeting with Kallstrom that properly should have included Dickinson. Even his friends acknowledge Bob Francis is an independent soul. Here he was clearly living up to his reputation.

About 1,110 FBI agents live and work in the vicinity of the crash. When the small team of NTSB investigators arrived at the scene, half of those agents were already there. Adding to the law enforcement presence were agents with the Bureau of Alcohol, Tobacco and Firearms, ATF.

Their presence made NTSB's Joshi wonder if somebody already knew what brought the plane down and had simply not told the safety board.

"This all showed me how devastating the accident was," Joshi recalled. "Eighty percent of the time, what we hear initially about a crash is wrong. When people say, 'It's a bomb,' I don't trust it. But their sheer presence was hard to set aside. All I could think when I saw the number of FBI and ATF agents was, 'What do they know that I don't know?'"

Peter Michael Santora owed a lot to the generosity of his next-door neighbor, Judith Yee. Santora claimed he'd won a half-million-dollar settlement in a lawsuit against a bank for misrepresenting his credit status. Yee had helped him invest the money and taught him how to follow his investments on the computer. She understood long-term financial planning. Santora respected that. It had sure paid off for her.

In her forties, Yee took severance rather than move when her employer, Mobil Oil, took its corporate headquarters from New York to Texas. She'd done so well on her own that she never bothered to look for another job as she'd originally planned. So Yee usually had time for Santora when he dropped by her apartment.

In the days following the crash, Santora offered to help sort out Yee's affairs, and her brother Ronald Yee gladly accepted, saying he couldn't get to New York from his home in Hawaii until the end of July, when he would have to take responsibility for both his sister's and his cousin Patricia Loo's estates because both were single and childless.

Santora got dental records to the medical examiner, located Judith's safety deposit box, and took care of other things using the key to her apartment and her mailbox, and the password to her computer and bank accounts.

Though a few of Judith's many friends warned Ronald about Santora, he was shocked to learn that Santora had taken advantage of his access. Before the year was over, Yee's estate would claim Santora had stolen more than $10,000 from the woman he maintained was his friend.

"Great luck," Merritt Birky thought, refolding the Long Island map and laying it on the passenger seat of his rental car. After chasing transportation accidents in some of the most remote locations around the world, the bearded chemist was happy to see the NTSB command post was going to be easy to get to.

Safety board investigators are assigned to a rotation of go-teams that respond immediately to public transportation accidents. The agency's only expert in fire and explosions, Birky was virtually always on call. He'd handled the ValuJet crash and had assisted the investigation into the explosion of the NASA shuttle *Challenger*.

With his sixtieth birthday coming up, he was less than five months from his retirement on the morning of July 18. He arrived at his NTSB office in Washington only to be told to get to Long Island right away. At the airports in Washington and Long Island, even at the Avis counter, the buzz had been that TWA Flight 800 had been shot down or bombed out of the sky. When Birky tuned the car radio to the local news, he heard more of the same from the FBI's Kallstrom.

"Something happened out there. We know that. If it was an act that was perpetrated by terrorists, we will find them. We will find the cowards."

Kallstrom's suggestion of terrorism annoyed Birky. The FBI was jumping to conclusions. If there was evidence the explosion was a criminal act, he had certainly had not heard about it. In Birky's experience, what was obvious one day could be obviously wrong the next. He put the broadcast out of his mind. As he headed east on the Long Island Expressway, he reviewed what he knew so far about the midair breakup of the 747.

There had been two explosions. The first, as the plane was beginning its ascent to 20,000 feet, seemed significant. He remembered other airline disasters in which bombs had been triggered by altitude-sensing devices. The second explosion was less curious. It had occurred some seconds later and could have been a consequence of the first blast.

He turned over various scenarios in his mind, and found himself thinking about a shipping accident he'd investigated years earlier. A partially filled grain storage container on a freighter had exploded when a spark of static ignited a volatile mix of air and grain dust.

The strength of the blast had impressed Birky. One-inch-thick steel walls had peeled away like banana skins. The scientist had come away with a new respect for what are called "low order" explosions, blasts created not by terrorists, but the laws of nature.

Birky had been told that Flight 800 left New York with its 12,890-gallon center fuel tank empty of all but fifty gallons of fuel. Were the vapors in the center tank enough to have turned it into a bomb?

He remembered the damage on the freighter's storage container and transferred the image to an aluminum fuel tank in the belly of a passenger jet two miles aloft. He had no doubt the destruction would be total.

In Washington, within hours of learning about the crash Federal Aviation Administration officials were in a lather over what they thought was pretty convincing evidence a missile had hit the jumbo jet. The clues had come in the form of air traffic control data from the FAA center responsible for flights within a sixty-mile radius of Kennedy Airport. Two of the controllers who spend twelve hours a day making sense of the luminous green sweeps and blips on round radar screens seemed to think they saw a target arching and then intersecting with TWA 800.

A manager for the FAA's air traffic control center had been ordered to bring the radar data to the FAA Technical Center in Atlantic City, New Jersey, for further analysis. He hurried to Atlantic City, where a playback of the data was turned into a videotape. Then what was seen on the tape was plotted on paper. Technicians in New Jersey faxed the paper to FAA headquarters in Washington, D.C.

All along, the New York controller had a disquieting feeling about his presence at the tech center. He could not have imagined that the information he'd brought with him from New York would be on its way to the White House situation room or that the fax to Washington would launch a conspiracy theory that would forever shadow the crash of TWA Flight 800.

8

One reason so many people assumed that a bomb or a missile had taken down TWA Flight 800 was the sheer incredibility of a jumbo jet simply exploding in flight.

In his 1990 book *The Final Call*, British aviation journalist Steven Barlay said, "There are no new types of air crashes—only people with short memories. Every accident," he writes, "happens either because somebody did not know where to draw the vital dividing line between the unforeseen and the unforeseeable or because well-meaning people deemed the risk acceptable."

For some of the investigators, TWA Flight 800 brought to mind an in-flight explosion aboard a Pan Am jet in 1963. The plane turned into a fireball at 5,000 feet and fell into a rain-sodden field, killing all eighty-one aboard. Yet the condition common to the Pan Am and TWA disasters, the practice of flying with fuel tanks in an explosive state, was deemed an acceptable risk for thirty-three years.

It was already dark and the rain was coming down steadily on the evening of December 8, 1963, as Pan American Flight 214, a Boeing 707, was readied for takeoff from Baltimore's Friendship International Airport. Fuel loaders pumped 12,000 pounds of kerosene-type fuel into the two large tanks on the plane's wings. The two outer wing tanks and the plane's center fuel tank were left as they were, with only puddles remaining of the more volatile Jet B aviation fuel loaded onto the jet earlier in the day in Puerto Rico. There was no need to refill those tanks for the short flight to Philadelphia, barely 100 miles away.

The trip would be the final leg of a full day of flying for the four-man flight crew. They had started early that morning in Philadelphia, stopping first in Baltimore and then continuing to San Juan.

From San Juan there were 142 passengers, sixty-nine of whom would get off at Baltimore. The other seventy-three passengers and the crew of eight would continue to Philadelphia.

Second Officer Paul Orringer, a former air force pilot, worked the radio on Flight 214. Orringer, thirty-two, was ready to finish his trip. At the request of his wife, Shelly, he had given up flying the glamorous routes to Bangkok, Johannesburg, and Rome to have more time at home. Orringer willingly agreed to fly domestically, an indication of just how domesticated he was becoming.

The passengers were as eager to land as Orringer. Among them were three sets of honeymooners, eighteen members of a Philadelphia country club returning from a men's-only golf outing, and two new mothers bringing infants to meet grandparents.

The flight should have been uneventful. The airline's safety record was good and the plane had been inspected the day before. Captain George Knuth, a Pan Am veteran and the safety officer for the pilots' union in New York, was in command. Baltimore to Philadelphia was a quick hop in a tight ship, with an experienced crew, on a well-traveled corridor. Weather was the only uncertainty.

A line of storms lined the flight path, creating turbulence, icing, and limited visibility. As Flight 214 approached Philadelphia at 8:42, the air traffic controller, Paul Alexy, offered the pilots a choice:

"I've got five aircraft, have elected to hold until . . . extreme winds have passed. Do you wish to be cleared for an approach or would you like to hold until the squall line passes Philadelphia?"

"We'll hold," came the reply.

Eight minutes later, at 8:50, Orringer told the controller that they were ready to approach the airport.

"No hurry," Orringer added, "just wanted you to know we'll accept a clearance." Still, nearly eight more minutes passed, during which Alexy alternated his attention between a column of planes in a similar situation and weather radar indicating no letup in the rain, clouds, and lightning.

Captain Knuth navigated the 707 in an elongated oval in the sky about forty-five miles southwest of Philadelphia, killing time, waiting for the storm to pass. A sharp crack startled the men on the flight deck. The plane began an abrupt turn to the left and the nose of the jet pitched down. An explosion in the fuel tank farthest out on the left wing had torn off twenty feet of the tip, nearly a third of its total length.

"Mayday, Mayday, Mayday, Clipper 214 out of control," Orringer's controlled but clearly frightened voice shot through the dark room of the Philadelphia control tower.

The 707 was dead weight falling. In seconds, it had lost half a mile of altitude. Captain Knuth in the left seat could see the lights on the ground coming up quickly toward the front of the plane.

Another call from Flight 214 grimly reported, "Here we go . . ." before the transmission ended midsentence.

Aided by the rushing wind and fed by highly flammable vapors, a stream of fire raced through the fuel vents and into the jetliner's center tank. An explosion there made a much louder sound. The pressure of the blast pushed out at the heart of the plane, tearing the structure and flipping the aircraft over.

The voice of the copilot of a National Airlines DC-8 circling just above where Pan Am Flight 214 had been holding now ricocheted through the tower, "Clipper 214 is going down in flames!"

Flight 214 hit a cornfield in Elkton, Maryland, upside down and flat. The impact tore a forty-foot crater in the ground, shattered windows in nearby homes, and knocked out electrical service to the farmhouse on the property. Clocks in the house stopped at 8:58 P.M.

Sometime after 3 A.M., Pan Am Captain Eugene Banning arrived at the crash scene. Banning, the Air Line Pilots Association's air safety investigator, had driven through the stormy night from his home in rural Connecticut.

Banning felt the loss of Flight 214 on several levels. He'd known Captain Knuth practically since the two men joined the airline in 1949. Ten years later, they shared the excitement of flying the North Atlantic routes in the Boeing 707, the world's newest airliner. Because both men were captains, Banning and Knuth had never flown together, but over the years Banning had flown with all the other crew members of Flight 214.

The crash was a blow on a professional level, too. Pan American pilots were proud to be flying for the first airline with jets. Until this night, Pan Am had perfect safety in the jet age; a five-year, 712-million-mile record of delivering eleven million passengers to their destinations without a single fatality.

Standing in the cold drizzle, Banning considered the smoldering remains of the 707 and the passengers who'd lost their lives on it. Bits of clothing and aircraft debris were snagged in the bare trees. It was a harsh scene illuminated by the headlights of police cruisers and the revolving red beacons of the fire trucks.

The 120-ton plane had hit the ground with tremendous force. Everything else about the disaster was indecipherable. "The best way to handle the sadness," Banning remembers thinking, was to "dig in and find out the cause."

To Raymond Gregg the cause of the crash was simple. Lightning had hit the plane. What Gregg and his wife, Joan, knew about air travel came primarily from watching the planes flying over their farm.

The couple had been watching television when a bolt of lightning flashed so brightly it drew Raymond to the window. It looked like dawn to Gregg.

On the morning of December 12, Banning and many other investigators didn't buy lightning as an explanation for the crash. In years of flying, Banning's planes had been struck by lightning with some frequency. Statistically, a plane can expect to be hit about every 3,000 hours of flight, or roughly once a year.

"We are not eliminating any possibilities," Civil Aeronautics Board* chairman Alan Boyd told reporters, adding that he had never heard of lightning destroying a plane in flight.

As it turned out, Gregg was right and the experts, Boyd and Banning among them, were wrong. But there was more to it than that.

The crash of Pan Am 214 was front-page headlines and not just because the crash itself was news. The disaster was also a wake-up call that the bigger the jetliner, the more deadly the consequences when something goes wrong.

As they began the routine of looking at past accidents, investigators were startled by a just-issued report from the National Aeronautics and Space Administration. After four years of study, the agency had determined that the 1959 explosion of a TWA Lockheed Constellation fifteen minutes after taking off from Malpensa Airport in Milan, Italy, had been caused by

---

*In addition to many other roles during the era of regulated commercial aviation, the Civil Aeronautics Board was responsible for investigating aviation accidents prior to the formation of the National Transportation Safety Board in 1967.

lightning igniting vapors in a nearly empty fuel tank on the right wing. Sixty-eight people died in the crash.

The conclusion, that lightning could be a serious threat to aviation, came on the very same day Pan Am 214 investigators recovered the left wing tip of the 707 a half mile away from the main wreckage. Several lightning strike marks were found on the wing skin along with a one-and-a-half-inch hole, encircled by fused molten metal.

The CAB had to rethink its earlier dismissal of lightning as the cause of the explosion. Even so, an unnamed source told the *New York Times* it was "a one in a million disaster."

It took a week for the CAB to conclude, tentatively, that lightning had set off an explosion of volatile vapors in Flight 214's wing tip fuel tank. An experienced crew might have been able to land the plane safely if that explosion had been the only one. The second, larger explosion in the center fuel tank doomed the plane and its passengers.

The 707 had been fueled in Puerto Rico with Jet B, a mix of kerosene and gasoline, primarily used by the air force, though it was used in commercial planes with some regularity at airports outside the United States. In the wing tanks, Jet B was added to the Jet A not consumed on the flight down from Baltimore. The center tank was filled almost entirely with Jet B.

The significant difference between the two fuels is their flash point, the temperature at which the vapors will ignite. For Jet B, ignition can take place between 0 and 50 degrees Fahrenheit. Jet A must be much hotter, between 100 and 170 degrees Fahrenheit to flame. Combining the two fuels increases the range of flammability. One expert characterizes the effect as bringing out the worst in both fuels.

Before 1963 ended, CAB safety bureau director Leon Tanguay wrote to the FAA, urging prohibition of the use of the more flammable Jet B, and asking for development of a method to prevent explosions by removing oxygen, an essential element of combustion, from airplane fuel tanks. This process is known as inerting.

Then, as now, determining the cause of a crash is crucial to avoid repeating the same mistakes. Accident probes often lead to new safety rules, a process morbidly called tombstone regulation.

Both Pan Am and TWA had already sustained deadly losses from fuel tank explosions when, in November 1964, another TWA disaster killed forty-nine people in Rome. By odd coincidence, this flight was also TWA Flight 800.

The TWA 707 had flown from New York to Paris, then to Rome, and was to continue to Athens. The plane never left the ground in Rome. Captain Vernon Lowell was accelerating down the runway when he noticed two engines were not functioning correctly and aborted the takeoff. The plane swerved and the right wing hit a truck resurfacing a nearby taxiway. The fuel spilling from the ruptured wing tank quickly ignited. Once again, fire raced through the fuel system vent lines, and an explosion tore through the center fuel tank, just as it happened over Raymond Gregg's Maryland farm eleven months earlier.

When the plane came to a stop, passengers in the rear of the plane were able to flee through the back exits, but not before seeing the people seated over the center fuel tank being tossed out of the plane with the explosion that blew the jetliner apart.

Pan Am 214 and TWA 800 in Rome were responsible for 130 deaths, but these were not enough for the FAA to mandate even the simplest of fixes.

Acting on its own, Pan Am discontinued using higher flammability Jet B. Two weeks later, TWA followed. The airlines acted because they thought the newspaper reports questioning the volatility of Jet B fuel would frighten passengers away. They made the decision despite an FAA special advisory group report concluding there was no clear-cut safety benefit for either fuel.

9

TWA made its first public statement on the loss of Flight 800 to Paris at 11 P.M. on July 17, 1996, more than three hours after the crash. Red-eyed and emotionally exhausted, Mike Kelly, vice president for airport operations, faced the cameras, trying to follow the advice of TWA's director of media relations, John McDonald, and "stick to the facts."

TWA had been in the news a good deal during the preceding decade: financial jams leading to wage concessions from the unions, the sale of airline assets, two bankruptcy filings, and six top-level leadership changes. Two years before, a runway collision between a TWA DC-9 and a general aviation airplane killed two people at Lambert Field in St. Louis.

This event was a bigger story than any of them. More than a thousand reporters left messages or paged McDonald in the first hours after the crash. He knew the airline had to get someone in New York to answer reporters' questions, and soon.

The problem was that TWA's vice president for corporate communications, Mark Abels, was overseas with chief executive officer Jeffrey Erickson. Erickson was in London trying to win back the landing and takeoff slots at Heathrow Airport that the airline's former owner, Carl Icahn, had sold for $470 million seven years earlier.

When Erickson went to TWA in 1994, it was with a mandate to right the company's teetering financial state. Resumption of service to London could go a long way toward achieving that goal. Marketing experts had predicted flights to Heathrow could become TWA's most profitable routes.

The trip had seemed successful. Three days of meetings with members of the British government had convinced Erickson that TWA would be given fair consideration amid the stiff competition from other American carriers. Just before midnight in London, Erickson and Abels went to their rooms for the night, optimistic about TWA's future. At 2:30 A.M., everything changed.

Mike Kelly had raced back to JFK from home after being called with news of the disaster. He did not take the time to shave or change clothes. He was still in chinos and tieless in a casual shirt when he was told he would serve as the airline's New York spokesman.

Kelly was informed that Erickson and Abels were trying to get to New York. McDonald explained from St. Louis that he couldn't leave TWA's corporate office.

McDonald quickly summed it up for Kelly: You're the ranking guy. You're gonna have to handle the press. I wouldn't ask you if it wasn't vital.

McDonald knew how competitive and aggressive New York reporters could be. He knew Kelly to be even-keeled and knowledgeable about the JFK operation. He would do fine until Abels and Erickson arrived the next morning.

Kelly was queasy at the thought of what had happened. He did not know which of his friends had been on the flight, nor did he know if TWA was somehow to blame for the accident.

Kelly began the news conference handicapped by a lack of information. The reporters were relentless. He took only those questions for which he had answers and wound up repeating himself.

Back in the tower, real information was coming in, but Kelly couldn't get back there. When he ended the news conference, a Port Authority official asked him if he'd mind sticking around a little bit longer. New York Mayor Rudolph Giuliani was on his way with a few questions of his own. Reluctantly, Kelly agreed.

About the same time, aboard the Suffolk County police boat *Romeo,* in the weird illumination of flares and flames, Vinnie Termine watched a fellow police officer vomiting over the rail. Thick kerosene fumes made Termine feel as if the inside of his nose and mouth were lined in oil. Particles of ash aggravated the burning in his eyes, and he worried he'd be the next one heaving.

They'd been working since 10, pulling bodies and body parts out of the slickened water and laying them on the deck of the twenty-five-foot twin-engine boat until a larger police boat came to take them to shore.

Termine tried not to look at the young girl who had been the first one retrieved by the officers. Seeing her floating on the water had shaken all four experienced police officers.

"It's like a war zone," Termine remembers thinking. But there were more victims bobbing around the boat, and there was no time to reflect on the horror around them.

So many of the victims had been torn apart that the officers quickly began to lose hope they'd see any survivors. Yet, ironically, it was the discovery of one perfectly intact victim, a uniformed TWA pilot, that confirmed that fear.

The man was fully dressed, right down to a dark navy blazer with gold wings on his breast pocket. He had no cuts, no visible damage; it looked as if he were sleeping. The fact that even he had not lived through whatever happened brought the officers to the consensus that nobody escaped.

. . .

Rudy Giuliani grows his thinning hair long on one side and brushes it over a wide furrow of scalp. His face is set in a perpetual bulldog scowl. A former federal prosecutor, he is often in scrappy in-your-face interrogator mode. When he greeted Kelly, it was with a soft voice and a matching handshake. He did not meet Kelly's eyes but gazed at the floor, telling Kelly he was there to help, and then he asked what happened.

So far, we know only that the plane disappeared from radar about eight thirty or so, Kelly replied. It was a flight to Paris. There were more than two hundred people aboard.

Still looking at the floor, Giuliani asked to see the passenger list.

We prepared a passenger list, Kelly said. It's being reconciled in St. Louis. I'll be happy to get it to you when it's released by the airline.

The mayor asked Kelly if it would be okay to visit the hotel on the airport property where the arriving families of Flight 800's passengers were being taken, a large Ramada by the highway.

The question caught Kelly off-guard. The mayor was asking if he would object? It didn't make sense.

Object? Of course not. It might be very helpful. Please go, he answered.

With that, Giuliani turned and started out of the building. Kelly felt a hand on his back. Turning around, he looked into the face of a woman who had arrived with the mayor. Though she did not identify herself, she asked that the list be prepared in alphabetical order, grouped by ZIP code, with phone numbers.

Kelly was flabbergasted; there was no way the airline would have immediate access to all that information or the ability to sort it out in that way, but We'll see was all he said.

Mary Anne Kelly and Jamie Hogan, cousins who worked together as hostesses at TWA's Ambassador Club at La Guardia

Airport, had dressed in business attire and headed to JFK after hearing the news bulletin about Flight 800. They entered TWA's main ticketing lobby a little after 10 P.M.

Since the airline's fall from its status as one of the nation's premier international carriers, the spacious lobby had started to look neglected. Tiles were missing from the steps of the wide curving staircases framing each side of the hall, paint was fading, and the large plate-glass windows to the airfield were foggy with age, but maintenance standards were high in the airy, crescent-shaped JFK Ambassador Club lounge, which sits like an elegant treehouse above the hall. The club is a restful, well-appointed waiting area for dues-paying frequent flyers. In the hours after the crash the club was set aside for families of Flight 800 passengers, though TWA had no idea what to do with them after that.

On their arrival at the club, Kelly and Hogan saw the receptionist engaged in what appeared to be an argument with several men. One of them turned to Kelly and asked if she could pull up a passenger manifest.

Kelly stepped around the crowd and over to the computer and started tapping a few keys. She still didn't know who the man was. Though she found his manner arrogant, she guessed he had the authority to be asking.

I'm sorry, it's locked, Kelly said after a few seconds, staring at the unresponsive screen. See, this information is totally locked up, none of us know how to get at it.

Well, where can I get information? the man said. I would like to know who, from TWA, is in charge?

Excuse me . . . Kelly replied, her own attitude now in full display.

Do you know who you are talking to? the man interrupted. I'm George Marlin, the head of the Port Authority, he said, the Port Authority of New York and New Jersey is the agency responsible for the operation of the area's three major airports.

Well, Mr. Marlin, I think you better go to the tower because we are not going to be able to help you here, she replied.

Still shaken by the confrontation, Kelly quickly called up to the TWA's operations tower to let Mike Kelly know that she and Hogan had arrived and were available to help. She also wanted to fill him in on what had just happened.

The Port Authority chief may not have been able to track Mike Kelly down, but someone from the agency had already laid out a plan for the TWA manager.

Mar, he said, the Port Authority has a bus waiting downstairs for the people who are in the club. I want you and Jamie to escort them. Get them safely to the Ramada.

Down a hallway about twenty feet from the reception desk was a spacious salon of sofas and club chairs arranged in several separate groupings. At the near end of the room was a television set tuned to CNN. Several people were watching as Hogan and Kelly stepped in.

Excuse me, I'm Jamie Hogan from TWA. If you'll come with me, we've set up a place for you at the airport hotel. The two women were standing just inside the entrance. Mary Anne and I can take you there now.

Wordlessly, the people in the room got up and followed the cousins out of the club, down the staircase, and onto the sidewalk in front of the terminal where the Port Authority bus was waiting.

In the absence of a decision from TWA, Marlin decided to send the family members to a hotel to get them away from the airport and the curiosity seekers and news reporters in the very public International Arrivals building.

The effort was not entirely successful. As the family members stepped onto the bus, half a dozen television crews on the pedestrian island twenty feet away spotted them, switched on 200-watt "sun guns," and rolled tape. Drivers had been instructed to keep the interior bus lights turned off, but the camera lights made any other illumination unnecessary. Intensely private scenes of despair were recorded and broadcast to the world.

For most of the five-minute drive, a middle-aged woman sobbed in the arms of her husband. As the bus pulled up in front of the hotel, Kelly moved over next to where the woman was seated and pressed into her hand a rosary she brought from home.

"Somebody's screwed up" was Ron Krick's first thought as he walked off the plane and saw Mike Stelzer on the jetway, alongside Captain Jim Cone. Krick and Cone had just flown into St. Louis from New Orleans. It was unheard of for Stelzer, a union rep in charge of pilot grievances, to be meeting pilots at their planes at 9 in the evening. The news he was bringing couldn't be good.

Greeting Krick, Stelzer asked the pilots to come with him.

What did we do wrong? Krick asked.

Nothing.

Okay, what did I do wrong?

Nothing. You didn't do anything wrong. I just want to talk to you for a minute, Stelzer replied.

Stelzer guided Krick out of the jetway while Cone followed behind. They passed the crowd of passengers waiting to board the plane to Louisville, and Stelzer led them to an empty waiting area nearby. Ron sat down but Cone remained standing. Krick was beginning to think Stelzer's attention was focused on him, not Cone, and it became obvious when Stelzer crouched down in front of Krick's chair and placed a hand on Ron's arm.

Ron, this is very difficult, Stelzer began, and I don't even know how to tell you this. Flight 800 has gone down and it's not thought that there are any survivors. Ollie is presumed lost.

Ollie? Krick said, confused. He wasn't on Flight 800. Ollie's trip is to Rome. He's on Flight 848.

For a moment, Krick felt relieved. He was not in trouble, and because he was certain that Oliver was not flying to Paris, the whole thing with Stelzer was just a big mistake.

Ron, he was, Stelzer said patiently. Flight 848 was canceled and Ollie was switched over to Flight 800. He paused. I'm sorry, Ron.

The news registered slowly. Ron had an image of a day twenty-two years before. He had the scare of his life watching three-year-old Oliver break away from his mother and race full throttle across a rural airfield in Monticello, Indiana, to greet his father taxiing in a single-engine prop plane. Horrified and helpless to act, Krick was certain his son would run into the spinning propeller and be killed. But even at three, Oliver had the good sense to respect the aircraft and stay clear of the blades.

A headlong enthusiasm for everything was Oliver Krick's hallmark. It was impossible to believe that he was dead.

Ron Krick, sagging into the airport chair, murmured repeatedly to himself, "No, no, no. It can't be true." Oliver, the Eagle Scout, the jokester, the scuba diver, was, above all, a lover of life, a survivor. He had to have survived this.

Margret Krick agreed. For days, she insisted her firstborn son had not been killed and was going to be found "hanging on to a piece of wreckage."

It took his arrival in a coffin on July 28 to break that hope.

"He used to fly in the cockpit. His final trip home he flew as cargo," she recalled. "That's how I realized he was gone."

Stelzer sat down next to Krick and put his arm over Krick's shoulder. Thirty minutes later, he drove him home.

1 0

The Ramada Plaza Hotel sits at the eastern edge of the seven-mile-square area that is John F. Kennedy International Airport. The modern and attractive hotel is often booked to capacity, but on the night of the crash of Flight 800 its landlord, the Port Authority of New York and New Jersey, was able to take over most of the 520 bedrooms as well as the hotel's grand ballroom and meeting rooms. The general manager turned his own suite of offices over to the mayor and governor of New York.

As family members arrived, uniformed TWA employees and a battalion of Port Authority police officers directed them through the lobby restaurant to the ballroom. Hotel staff had set up tables and chairs and placed sandwiches, coffee, and drinks on a long table by the wall.

There was no television in the ballroom. People in touch with the outside world through their cell phones gave the latest news, which was then passed along from table to table. The most relevant information, the list of passengers on Flight 800, was unavailable until after midnight when Marlin arrived.

The Port Authority is responsible for running the bridges, tunnels, and ports in the New York/New Jersey area, along with three major airports: Newark, La Guardia, and JFK. Kennedy is not the busiest airport in the country, though it moves an impressive thirty-one million people a year, more than half of them international travelers. The Port Authority is a political agency controlling a $2.5 billion annual budget, operating from offices in the famous twin towers of the World Trade Center.

Marlin, the Port Authority's forceful director at the time, was an appointee of New York Governor George Pataki, a Republican whose antipathy for his fellow Republican Rudy Giuliani goes back to Giuliani's decision not to endorse Pataki in the 1994 election. Marlin is no great friend of the mayor, either, having run against him as a Conservative Party candidate for mayor.

At the Ramada, however, Marlin and Giuliani put aside their political differences and united in their common outrage that TWA had left the families without information about the crash or the names of the passengers on board.

With a Port Authority police force of 300 and staff at JFK in the thousands, Marlin was able to get his hands on a passenger list for Flight 800 without TWA's help. The document was one of several printed in the hours before a flight departs to help flight attendants count special meals and accommodate individual requests. Once the doors to the plane are closed, these "spill" lists are discarded and a new, final list is generated.

Marlin quickly admitted that his list was not official. It was not completely accurate, either. Two passengers had switched off the flight at the last minute. To the delight and relief of their families, the men arrived in Rome on Alitalia's Flight 611 the morning of July 18.

Entering the ballroom, Marlin could see Giuliani mingling in the crowd, embracing some people, nodding sympathetically with others. Pulling him away from a group, Marlin explained in a low whisper that he had a passenger list in his breast pocket.

I can't tell you how accurate it is, Marlin remembers warning the mayor, but we have a list which is more than these nitwits from TWA have. What do you want to do?

The mayor paused to think. For some time now he'd been listening to the same questions over and over again: What had happened? Were there any survivors? Why hadn't TWA issued some kind of statement? When would someone answer their questions? Queries for which the mayor had no answers. He had questions, too. Giuliani was a personal friend of Kurt Rhein, the Connecticut businessman traveling with Hank Gray. The mayor didn't even have information for Rhein's wife.

Acting on years of experience as an advocate with a highly developed sense of righteousness, Giuliani accepted the list from Marlin, cleared his throat, and told the families he was going to help them.

Mary Anne Kelly thought that Giuliani was taking political advantage of the crisis and the scene disgusted her. She'd been on the phone half a dozen times, begging for a more senior TWA staffer to come down and do something, say something. Higher up on the corporate ladder, Mike Kelly was doing the same thing.

In phone conversations with TWA officials John McDonald in St. Louis and Mark Abels aboard the corporate jet carrying CEO Jeffrey Erickson back to the States, Mike Kelly pleaded, We've got to get an official passenger list. Mayor Giuliani is on my ass. The girls in the Ramada say the families want somebody from the airline to provide answers. It's not good for us not to have somebody there.

McDonald and Abels said they'd spoken with Giuliani, too. He'd told both of them, quite vigorously, that he wanted the company to provide him with an official list even though TWA was still verifying the names.

You are absolutely not, under any circumstances to go down to the Ramada, Kelly was told. Don't go near it. There is nothing you can do but piss people off.

"There's no upside and quite a lot of downside," McDonald remembers advising.

Shortly after 2 A.M., in a second telephone call from the jet carrying the TWA executives, Kelly was thanked for his efforts and told to go home for the night. Richard Roberts, a TWA vice president who had arrived from Washington, D.C., would take over on the crash.

Going home may have seemed reasonable to Kelly's superiors, but Kelly was incredulous at the order. He left Hangar 12 and went to his office in the airport terminal to start returning the calls on his pager. At 3:30, he finally did as he was told. He drove home and spent the next two hours trying to sleep.

For all the rest he got, it would have been better for TWA had Kelly not gone home at all. Mayor Giuliani stayed at the airport through the night, and when he learned Kelly had not, he tore into the airline for crass insensitivity to the suffering of the families. By dawn, they had neither seen an official passenger list nor heard from any executives of TWA.

Giuliani left the airport early in the morning to appear on three morning news programs and used each of them as a forum for blasting Mike Kelly for disappearing and TWA for abandoning the families.

George Marlin was not an easy man to track down the night of the crash. When his wife finally got through to him, she was brief and to the point: Call Joe.

Marlin's old friend, Joseph Darden, a bond trader living in Houston, had sent a fax to Marlin's home asking him to help a neighbor who had lost his wife and daughters in the crash. That would have been enough for Marlin, whose affection for Darden is deep and who can be compassionate despite a gruff and impatient nature. Before he could respond, he was asked to take a call from his former campaign manager, Eileen Long, who was working in the governor's press office.

George, Long said, right to the point, I've just hung up the phone with a guy from Texas. It's the saddest thing. His wife and daughters were supposed to be on Flight 800, but he can't get through to TWA. Do you have a way to confirm the passenger list?

I've been working on it all night, Marlin replied. Weird thing is, I think this is the same guy my friend Joe Darden is calling me about. Give me some time. I think we can help him.

The first thing Jeffrey Erickson and Mark Abels planned to do on arriving at JFK Thursday morning was get the latest news on the disaster so that they could be prepared for the news conference the airline had scheduled for 11 A.M.

Journalists had been arriving at the airport by the dozens, augmenting the rumpled, irritable ones who had been at the airport through the night, trying to ferret out fresh scraps of news for eager editors back at the office. Technicians from all the major television and radio networks were running cables, positioning microphones, and setting up lights so that the event could be broadcast as it happened.

From the moment Erickson set foot inside TWA's Hangar 12, his ear was pressed to a telephone taking calls from President Clinton, the administrator of the FAA, the Department of Transportation, and the U.S. ambassador to France, Pamela Harriman. He had no time to be briefed for the news conference, so TWA's top man made a quick statement of condolence to the families and an equally brief plea for understanding, explaining that the company had lost many of its own employees.

The message was considered by many to be too short, the messenger too wooden. Sixteen hours had passed since the accident, relatives had still not been notified, and TWA's leader was looking for sympathy rather than forgiveness. TWA couldn't do anything right.

Appearances to the contrary, TWA had a well-thought-out disaster plan. Its crisis team included special health services

director Johanna O'Flaherty, who had been baptized at Pan
Am, when she helped families of the passengers killed in the
bombing of Flight 103 over Scotland in 1988. In 1995,
O'Flaherty and a dozen other TWA staffers spent time in
Washington with U.S. Army Colonel Michael Spinello, learning
about disaster planning and notification of next of kin.

"Nobody does it better than the military," Spinello boasts
unabashedly. His term as deputy chief of staff for personnel of
the Army spanned the first six years of the nineties, during
which thousands of military personnel were killed. In each case
the next of kin was notified in person. "Someone is there in a
dress uniform, an officer and a chaplain, and if there's any
knowledge of health problems in the family, a medical person
might be there, too."

Spinello, a tall, intensely garrulous West Point graduate,
spent a day teaching O'Flaherty and the others what years of
experience had taught him about disasters, victims, and notification of survivors.

It was all fascinating to O'Flaherty, who is known at TWA's
corporate offices, with some affection, as the "Disaster Queen."
Still, notification of next of kin is not her job. That task is
assigned to the airline's reservations department, which has a
large number of employees and phone lines and access to passenger information. O'Flaherty is limited to developing and
maintaining an effective corps of employees trained to attend
to the needs of survivors. In this area she was unquestionably
successful. During the weeks following the crash, her family
escort program was the only TWA effort praised unilaterally.

That TWA was the only airline of more than a dozen invited
to accept Spinello's offer of training was indicative of
O'Flaherty's foresight. She knew what her colleagues did not
seem to realize: Policy makers were already dissatisfied with the
way airlines were handling this delicate aspect of the business.

The issue of airline disaster response came to Jim Hall's
attention in early 1995 the night before the public hearing into

the crash of USAir 427, which went down in Pittsburgh in September 1994. It was the airline's fifth crash in as many years. "USAir was so shell-shocked, they weren't handling it very well," said the NTSB's Goelz.

That night, Goelz and Hall spent three hours with the families, who were angered by what they considered the insensitive treatment of the airline and a lack of information from the NTSB.

"There were a lot of people on my staff who said I should not meet with them, that we needed to keep the accident investigation separated from the families' emotions," Hall explained. "My response then, as it is now, is if we're not doing these investigations for the family members, who in the hell are we doing them for?" Crash detectives wince at Hall's question, knowing, as Hall surely does, that investigations are conducted to protect future passengers. But the chairman's sentiment spoke to a new reality. Victims' families were becoming a political force.

At the meeting with Hall the Pittsburgh crash families wanted to view the wreckage of the 737 even though the plane was mostly pulverized. They had not been given access, even though the news media had.

"It's not pretty," noted aviation attorney Arthur Wolk, who represented the families of five USAir passengers. "By the time you put it together in some form, at least they could get an idea as to where their loved ones had been at the end, and that can be helpful to begin closure."

It was a big change for the scientifically minded safety board even to consider the subjective and emotionally charged concerns of the families. Goelz recalled that the visit to the hangar was "the turning point."

Hall's departure from safety board precedent does not surprise those who know him. Estee Harris, who worked with Hall in Tennessee politics for years, claims Hall has "the ability to see around corners." Hall saw clearly that victim families could no longer be ignored in the aftermath of an aviation disaster.

He assigned to Goelz, his most trusted aide, the task of coming up with a policy to meet their needs.

The most radical proposal would have removed airlines entirely from any direct involvement with families. There was also discussion about whether a third party should be brought in for the notification, counseling, and return of human remains and personal effects to the survivors.

An Aviation Disaster Family Assistance Act was proposed to the Air Transport Association, an airline industry group, and a plan was introduced in Congress. Action in either arena would have been slow in coming had TWA not stumbled so publicly following the crash of Flight 800. In September 1996, President Clinton signed an executive order making the NTSB responsible for handling the needs of family members following aviation disasters. In October, Congress made it law.

As it turned out, O'Flaherty was the first TWA executive to take charge at the Ramada. Upon her arrival, she assigned TWA volunteers who had been arriving throughout the night to families who needed them. These specially trained employees would run errands or provide any assistance needed.

O'Flaherty had been vacationing in Santa Barbara when she was called with news of the disaster. Hopping the red-eye to New York, she arrived at JFK at dawn the day after the crash.

O'Flaherty's career began when she left Ireland to work as a Pan American stewardess in 1970. It was such a daring and glamorous move that the local paper ran it as a news story. Good looks and an affinity for people got her involved in flying. An interest in what went on inside their heads turned her job into a vocation. O'Flaherty worked and attended college, obtained a master's in clinical psychology, and turned her attention to counseling.

By the time of her arrival at the Ramada, the charming mannerisms remained of the former flight attendant who'd traveled so far from home seeking adventure. The woman who took the microphone at the hotel ballroom had an understand-

ing of what her audience was going through and a steely resolve
to help them.

The first thing she said to the families was simple. She told
them, "I'm sorry."

Two hundred and thirty people had died on a TWA flight
and it had taken the airline ten hours to actually get an executive on the scene to make apologies. Even O'Flaherty couldn't
negate that.

Joseph Lychner endured hours of uncertainty punctuated by
periods of intense frustration when he actually got through on
TWA's 800 number. Lychner, a successful entrepreneur in
Houston, was not a man to let roadblocks stop him. Unable to
get anywhere with the airline, he'd called the New York governor's office in Albany, where Pataki's press secretary, Eileen
Long, picked up the phone around midnight. After listening to
him for about five minutes, Long promised to do what she
could to get information for him.

Lychner's wife, Pam, had worked for TWA as a flight attendant on and off for a dozen years starting in 1978. She loved
the free travel, though flying standby was complicated when
flying with her daughters, nine-year-old Shannon and six-year-
old Kate. She was pleased that she and the girls were able to get
seats together in the forty-fourth row one hour before Flight
800 was scheduled to depart. She called home and left a message for Joe on the answering machine that she and the girls
were on their way to Paris.

When Long called Lychner back shortly after 2 in the morning, he was somewhat prepared.

Mr. Lychner, I'm not even sure how to say this, Long began,
and I'm sorry I don't know what to say. I've been able to obtain
a passenger list for Flight 800. It's not official. I'm really sorry,
she stammered, but Pam and Shannon and Kate, their names
are on the list. They were on the flight.

A long silence followed from Lychner's side of the line. Eileen was just about to say something when she heard him sigh, followed by, I know.

She was glad she hadn't interrupted the silence. Still, she felt powerless to lessen his grief. She decided then and there she was going to do something for Joe Lychner.

1 1

Had it not been such a horrible disaster that brought the federal agencies to Long Island, the contrast between the FBI and the NTSB's contribution of personnel would have been comical.

The NTSB had sent ten engineers and scientists armed with little more than pencils and clipboards. They would beg or borrow anything they needed as was their custom.

At their first meeting the morning following the crash, Kallstrom explained to Francis what had been done so far and asked what the FBI could do to help the safety board.

"He said he needed a place to work and a phone," said Kallstrom's chief deputy, Tom Pickard, who was in the meeting. Pickard says he and Kallstrom were disbelieving. "You don't have a phone?" they thought.

The FBI had rolled an eighteen-foot mobile command post onto the Coast Guard property, an air-conditioned workspace complete with power, communications, and computers. Again, Pickard remembers Francis remarking on the setup and saying, I need a trailer like you've got. In an act of what can be characteristic generosity, Kallstrom turned to Pickard and said, Tommie, get him one.

Pickard was still wondering how he was going to carry out his boss's hastily assigned order as the men were walking out of the Coast Guard building. Just then, a second FBI van, a similarly equipped Winnebago, was coming down the driveway. Pickard pointed it out to Francis and said, Bob, there's your trailer.

The FBI wound up providing the safety board workers with cars and cell phones, too. As freehanded as Kallstrom was, he and his top deputies were privately concerned about just what the NTSB would be able to offer.

"With all we'd heard about the NTSB, how they are the premier crash investigation agency in the world," Pickard explains, "Kallstrom and I expected them to come with all sorts of stuff and a lot of people. We were surprised when Francis introduced his staff. We thought, 'This is it?'"

This disparity was not lost on the members of the NTSB go-team. Several suggested it reinforced the sense that they were merely there for the moment, that the cops swarming over the scene would get the bad guy, make a case, and wrap things up. The one area where no crime fighter could replace them was on the important job of retrieving and reading the information on the 747's black boxes.

A New York City diving squad boat had been offered. After the meeting, sir, we can go, an eager officer had volunteered during one of the first meetings in a packed room on an upper floor of the Coast Guard station; We have room for two.

The police offered a hydrophone, a portable underwater listening device, to Deepak Joshi and Frank Hilldrup. The receiver is submerged in the water by someone wearing headsets who then listens for the pings being emitted from the boxes.

Ice chests were also put on the boat, to transport the recorders. Keeping them immersed in sea water would prevent corrosion of the data until the magnetic tapes were back in Washington ready to be examined in the safety board's lab.

Perhaps it was the still, glassy Atlantic promising smooth sailing, or the crystal-clear summer sky, but whatever the reason, the men left the inlet that afternoon optimistic they'd soon have what they were after. The search began in the area where debris had been floating, but pass after pass was unsuccessful. After a few hours, the speedboat returned to shore. A new plan and a longer commitment of time would be necessary.

The bodies retrieved from the Atlantic the night of the crash remained in the refrigerated trucks that had carried them to the Suffolk County Medical Examiner's Office in Hauppauge, Long Island, a forty-five-minute drive from the Coast Guard station in East Moriches. The refrigeration generators created a soft, steady hum in the shady alcove of the loading dock.

In his office two floors above, Dr. Charles Wetli was on the phone with Dr. Mary Hibberd, then-director of the county's Health Department. Staffers from her office had been asked to arrange for the families of passengers to be bussed out to Suffolk County and settled in a handful of hotels closer to the medical examiner's office.

Having the families nearby was part of the county's disaster plan, based on a fictional scenario in which, coincidentally, a 747 crashes at the county airport killing 250 passengers and crew. Dr. Wetli, who helped write the plan, explained, "We wanted them close to the government building so we could show them photos, jewelry, or other personal artifacts and have daily briefings with the families."

But Suffolk County's plan did not fit the plans already made by New York City. The director of the Mayor's Office of Emergency Services notified Suffolk officials that the grief-stricken relatives, now numbering in the hundreds, would remain in New York City, where Mayor Rudy Giuliani had assumed the role of family advocate.

Giuliani detractors, many of them in Suffolk County, grumbled among themselves that Giuliani was staking his claim with the families, having missed out on jurisdiction over the area's biggest disaster since the World Trade Center bombing. Still, there were legitimate reasons to keep the family members, who soon swelled to 600, where they were.

"Two in this motel, seven in that motel, scattered all over Suffolk?" asked the chief operating officer of the Port Authority, David Feeley, raising the concern of those participating in the discussion. "We had them fairly well contained and they were getting all the information as fast as it was available."

Suffolk County officials didn't agree. Dr. Wetli was worried.

Keeping the families so far away is going to cause problems. It's ridiculous. These people need to be nearby so that we can get to them when we need them, Wetli told Hibberd.

Wetli could not have imagined the army of city, airline, law enforcement, and social service workers already camped at the hotel. Tables for various agencies lined the hallway to the ballroom, a we're-here-to-help-you bazaar for the families to pass en route to the regular briefings. Packing it all up now was out of the question.

The Suffolk County executives could foresee logistical problems, though none expected that the combination of geography and politics would result in lasting hostilities. Wetli was experienced enough to recognize an inauspicious beginning. He slammed down the telephone in frustration.

Mike Kelly watched the Erickson/Abels news conference from the back of the room. He was now showered, shaved, and suited up for the day and was to serve as Erickson's escort. At the end of the news conference, an aide to Mayor Giuliani notified Kelly that the mayor and the governor wanted to see Erickson at the office they were using at the Ramada.

The three men went to the smoky, crowded conference room filled with uniformed police officers, staffers for Mayor Giuliani, New York Governor Pataki, city police commissioner

Howard Safir, and George Marlin. Giuliani leaped to his feet at the arrival of Erickson and Abels. Kelly stayed outside the office and the door was closed. The politicians vented the frustrations that had been building over the past sixteen hours at TWA's utter lack of a plan. Erickson noted wryly to himself that the TWA contingent was the only one that had come to the meeting without armed escorts.

The hours between the time Erickson and Abels entered the office and the time they emerged were spent in a frenetic exchange of telephone calls with TWA executives in St. Louis, trying to get a complete passenger list. When, around 2 P.M., an official list of those aboard Flight 800 was faxed to the Ramada, George Marlin says it was with the admonition from Abels that he was not sure of about sixteen names, the high school French Club students from Pennsylvania.

There are varying explanations as to why this official list was wrong, none of which had any effect on the politicians. Every additional minute spent waiting for a final and accurate list pushed them into a deeper state of fury.

It was two more hours before there was unanimous agreement on the list. Mike Kelly noted with a silent vindication that the official list and the one he'd shipped to St. Louis at 11 the night before were identical.

Giuliani and Pataki left the hotel and went directly to the area where reporters were waiting outside. The men announced themselves satisfied, with a reminder from Giuliani that TWA had been "woefully inadequate" in its response.

By 4 in the afternoon in Paris, TWA station manager Patrice Soualle walked quickly down the dimly lit hallway, passed the lockers where a few office workers were having a smoke by the vending machines, and pushed open the door to the large seminar hall on the sixth floor of the administration building. He went no farther than the small windowless anteroom.

The crisis center where he'd passed the day had been noisy and busy. Here it was silent. Crisis volunteers sat at a registration table directly in front of him. Off to the side there was a steady hum of quiet conversation layered with occasional sobs from some of the thirty-five people clustered around small tables.

He approached the reception table and asked the three young women sitting there, How are you doing?

All three nodded, murmuring that they were okay. Soualle, the station manager, unconsciously put his hand over his jacket pocket and said, I have the list here and I am going to inform the families.

Pulling the list out of his pocket, he asked one of the women to get the family of the first passenger named.

He said, I intend to speak to them here for the utmost privacy. I will tell them in the most gentle way, but it is going to be emotional as you can imagine.

You two, when I finish, be ready to escort those families that do not have a crisis volunteer to wherever they need to go. You've probably been told, but let me remind you that there is a room for the family to rest. You should arrange transportation for them to the United States if they wish, or back to their homes.

Soualle tried to give each of the women a reassuring smile. But at his most relaxed, the diminutive station manager was somewhat rigid. And by now, he'd spent thirteen hours moving from one smoke-filled crisis room to another, snapping orders to some, taking orders from others, with no time rest or eat. Now he faced the most difficult task of the day.

"I told each family, as they came up, 'We have the final list and I have to inform you that your loved one is on the list,'" Soualle recalled.

"Some people cried. No one embraced me. In fact, I found it odd, there was no anger, no one asked how it happened. After the confirmation, frankly, they were no longer even listening. They were in another world."

. . .

The FBI's Jim Kallstrom was eager to understand air crash investigations. It was one reason he was on this forty-minute boat trip to the scene of the accident with Bob Francis. Both men wanted to work out a way to handle this case together.

Aboard the 225-foot Coast Guard cutter *Juniper*, Kallstrom listened as Francis described the cooperative nature of safety board investigations. We form groups, he told Kallstrom, involving the airplane manufacturer, the airline, the people who work for the airline, and the manufacturers of any equipment that may have contributed to the accident.

Francis was a bit surprised that Kallstrom didn't know this already. Sure, compared to the FBI, the safety board was a humble, little-known agency, an independent executive branch office that employed fewer than 400 people in 1996. Still, he'd been on the safety board for just a year, and he'd already worked with the FBI on two suspicious accidents: the ValuJet crash in May that turned out not to be an act of intentional sabotage; and a train derailment in Arizona in the fall of 1995 that was. Someone had damaged a section of the railroad track, and four cars of the Sunset Limited plunged off a thirty-foot trestle into a ravine, killing a twenty-year Amtrak employee and injuring scores of others.

Standing on the *Juniper*, on the morning of July 18, the two men looked out onto a sea littered with detritus: seat cushions, life jackets, chunks of unidentifiable aircraft parts. It was nothing like the ValuJet crash, a horribly violent event that left little wreckage above the surface of the swamp in which it fell. Francis's brief experience at the NTSB told him crime fighters are often present for the first days after a transportation accident. In the Amtrak case, he'd stayed on for a few days after the FBI took over the probe. "We were doing press stuff together, supporting their investigation with expertise that the FBI does not have," Francis recalled.

Kallstrom could understand the two government agencies working together, but he was apprehensive about bringing into the team the aircraft manufacturer and others with an interest in the outcome.

Kallstrom became deputy director of the FBI at the same time Francis joined the NTSB, his promotion the result of twenty-eight years of successful crime fighting. He kept his own counsel, never blabbing to others what he was working on. As he listened to Francis's plans to collaborate, an incredulous "How could that work?" was screaming through his head.

Full of confidence and ambition, Kallstrom was facing the biggest case of his career. No partnership role with the safety board and half a dozen other groups was going to jeopardize it.

As it turned out, Kallstrom did not find his "Eureka" piece, hundreds of agents did not reach "critical mass," and Flight 800 became the first accident where the issue of which agency would be in charge was so hotly debated that the White House had to intervene.

"My job was to hose people down and say, 'Back to your corners,'" cabinet secretary Kitty Higgins said, recounting the many times the feuding participants ended up bringing their complaints to her office. She found herself reminding them, "This is about finding out what happened."

Though the official word was that the FBI and NTSB were conducting parallel and cooperative investigations, almost from the start, Jim Hall, the thin-skinned chairman of the NTSB, was offended by what he saw as a lack of respect for his agency's independent authority.

"There was really a lack of coordination both at the level of the Secretary of Transportation and the FBI in respecting the role the NTSB had in this particular matter," Hall said. He accused his own agency's Bob Francis of going beyond his authority. "There were certain meetings that he had with Mr. Kallstrom independent of the investigative staff," Hall continued, "in which he made decisions impacting the investigation."

Back at headquarters, it was felt that Francis and Kallstrom's symbiotic relationship was disproportionally beneficial to the FBI.

Francis dismissed the controversy by saying, "This is a Washington hang-up, that one agency has to be in charge."

President Clinton appointed Francis to the vice chairmanship of the safety board in 1995. His job, along with the other three board members, is to hear testimony and determine probable cause in transportation accidents and to offer safety recommendations for government regulators. As a board member, he does not report to the chairman, and no safety board employees outside of his own office staff report to him. This odd chain of authority baffled Kallstrom.

"I don't care what the book says, Francis was in charge," says Kallstrom. He was the one in the meetings with myself and the navy. He was the one making the decisions that had to be made. So whether he was technically in charge or not, he was in charge at the scene."

Francis says he was constantly directing authority to Dickinson, the investigator in charge, but it was Francis's face on television virtually twenty-four hours a day in the early days, leading the public, the press, and even Kallstrom to think otherwise. After all, Francis, a licensed pilot, was not inexperienced in aviation matters. He'd joined the safety board after twenty years at the FAA.

"Bob probably thought he was in charge," an old friend of his theorized, to which Francis replied, "I understood and I understand now that I was not in charge" of the investigation.

Confusion about who was in charge was not confined to the top leaders of the primary agencies. The most outrageous example was an impostor who was calling the shots at the East Moriches Coast Guard Station helipad for three days before he was discovered. Around 10 P.M. the night of the crash, David Williams, thirty, presented himself to Suffolk County Police Officer Mike Ryan at the East Moriches Fire Station. A few public information officers, Ryan among them, stood in the midst of a jostling crowd of journalists who were trying to get to the scene. The officers were looking for a way to accommodate reporters yet keep them out of the way of the rescue operation. In the middle of the confusion, a young man pushed his way closer to Ryan.

Sir, the young man said, I happened to be in the area and I'm experienced with this type of activity. Perhaps I can be of assistance?

Ryan assessed the man standing next to him, clean-shaven and crisply dressed in an olive flight suit and a red cravat, and wearing the rank of a lieutenant colonel.

Gesturing over his shoulder Ryan replied, Present yourself to the fire commander. I'm sure they could use you.

The exchange was so quick that had Ryan never seen David Williams again, he probably would have forgotten all about it. But not much later, he did see him, at the Coast Guard station, and his presence couldn't be missed.

Williams was still wearing his flight suit and he was using light wands to direct helicopters taking off and landing at the small patty of tar that served as a helipad. It was curious to a number of people that such a high-ranking Army officer was pitching in, doing such a dirty, noisy job.

Williams, large enough to be imposing were he not so ingratiatingly friendly, had been introduced to Elmo Peters, the Coast Guard commander working that night. After telling Peters he was trained to direct helicopter traffic, he was given officer quarters, meals, and use of the telephone, and put to work.

To some, Williams seemed to know what he was doing. When FBI agent Bob Knapp needed the station coordinates so that Jim Kallstrom's helicopter could land at dawn the following morning, he turned to Williams for the information. The helicopter landed without incident. Police helicopter pilots do not usually depend on instructions from the ground, since they are solely responsible for the safe landing of their aircraft.

"In the military, it's customary to have people to do that. And they train with those people. But in the civilian arena, you're on your own," explained New York State Police helicopter pilot Thomas Corrigan, who made many flights into the Coast Guard station at East Moriches after the crash.

Although Williams never served in the army, he had a working knowledge of helicopter approach and takeoff procedures. Some of his more unusual signals were the subject of discussion among the pilots, but few wanted to question a man two ranks below general.

"I saw his rank and I wasn't afraid to tell him he was screwing up, but I was aware of the fact that he was a lieutenant colonel," said Coast Guard pilot John Knotts, the aviation safety officer for the Coast Guard station at Brooklyn.

"At one point he was trying to land an H-60 in a very small place while I'm trying to load [New Jersey] Senator [Frank] Lautenberg on my aircraft to do an overfly. It was totally unsafe, and I went and said something to group personnel."

Chris Baur had a similar experience on July 20. He had a close call while trying to land at the station because people had wandered into the landing zone. Now, as he waited for passengers to board, he heard Williams instructing another helicopter into the small clearing and motioning for him to take off.

He wants us to take off, Baur's copilot Rodney Lisec said, in what was more of a question than a statement.

Ignore that man. He's a moron, Baur replied with no hesitation at all. We're flying, not him, and we're not moving till we get everybody in.

When the Pave Hawk lifted off and cleared the area, Baur got on the radio with a question of his own.

What the hell is this guy doing?

The army is running the landing zone, came the reply, and this guy is in charge.

He's not safe, Baur announced, and there's gonna be an accident. Call somebody and get this guy out of here."

Baur and the unidentified pilot on the radio didn't know that Williams had been asked to put down the wands and stay out of the way two days earlier.

Colonel Frank Intini of the Army Aviation Group had noticed that Williams was wearing army insignia and medals on an air force flight suit and gotten suspicious. For some reason, no other action was taken, and Williams remained at the Coast Guard station.

It wasn't until Intini saw Williams back in action at the landing zone on Saturday that police were asked to take over. They questioned Williams and then escorted him off the property.

How many helicopters had Williams assisted in landing at the command post? "Everybody's," laughed the FBI's Knapp, thinking about it later.

. . .

Adrian Fassett, Williams's boss, was not amused. After reading about the escapade in the local newspaper, he wondered what else was false about the man he'd hired a year earlier to work as a program director for the handicapped clients of his nonprofit agency.

His search for the whole story ultimately resulted in Williams's conviction on state and federal fraud charges. For at least three years, Williams had impersonated a military officer, sleeping in base housing, flying on military aircraft, and even conducting training and inspections at the Aeromedical Evacuation Squadron at McGuire Air Force Base in New Jersey. Williams had apparently created orders by lifting the documents from a friend's computer disk.

When questioned by a reporter from a Long Island newspaper, Art Covello, a spokesman for McGuire AFB, responded, "Are we embarrassed? No. This guy came with credentials. His face became known."

Fassett's calls to state authorities prompted an investigation into Williams's claims of being a physician, his practice of writing prescriptions for friends, and even signing off on medical disability certifications.

"In his program, people with retarded or developmental disabilities have to be diagnosed," explains Fassett. "Now what we're supposed to do is take our clients to a doctor to certify the disability. We found out after I fired David that he had been signing off on it as doctor. "

There was no reason for Williams to do such a thing. The medical evaluations were free to clients, and Williams did not make any money by signing the forms himself.

It is possible that none of what was learned about Williams would have been discovered at all had Fassett not been driven by his own curiosity. The police at the Coast Guard station simply escorted him off the property and let him walk away.

The law came down a lot harder on Tonice Sgrignoli.

. . .

Driving to Kennedy Airport the Saturday following the crash, Sgrignoli, forty-three, was both excited and apprehensive. This story was big, bigger than anything she'd done since the *New York Post* gave her the opportunity to leave the copy desk and work as a reporter on weekends. What a relief from the nonsense they were usually throwing her way.

"Tonice, how 'bout you try to sneak onto the governor's estate and see if the alarm works? Maybe taxpayers don't have to spend $50,000 for a new security system."

"Tonice, go grab the mayor at the Holocaust Memorial and get his reaction to the new *Sunday Post*."

Her assignment today was to report on what was going on with the families of the victims. If she handled it right, this story could get her off the copy desk for good.

At the TWA International Arrivals area, Sgrignoli congratulated herself for not wasting her time at the press camp outside the Ramada Hotel. At the airport, she'd run into a *Post* reporter who'd been called back to work from his vacation in Paris. He was happy to give her the boarding card he'd used for TWA's Flight 925 from Paris. With that she walked out to the gate, where she hoped to find arriving family members to interview.

The gate was empty when she got there, though. She didn't have time to plan her next move before a tall, graying man with model good looks and a TWA ID clipped to his shirt was standing over her, asking if she needed help. It was Michael Kempen, the flight attendant with the airline's family escort program.

I'm here to meet someone who lost family on Flight 800, she explained.

No problem, he answered, in a manner that was authoritative and reassuring. They've been taken to the Ramada. I'm on my way there now. Come, I'll take you.

As they walked through the long tunnel-like corridor, her mind raced. Kempen assumed she was family. How far would he take her? Would she have to lie when she got to the Ramada? Could she make up a story? What would she say?

Entering the main TWA concourse, Sgrignoli, an attractive blond in a fuchsia sundress, felt conspicuous. She prayed the *Post* photographer and reporter were gone, or at least that they would not give her away. She needed time to think. Kempen had interpreted her preoccupation as a desire to be left alone, and the two walked in silence all the way to the Port Authority car waiting at the curb.

Kempen held the door for Sgrignoli, then followed her into the backseat of the black sedan. He pulled the door closed behind them. The *Post* photographer, who had spotted Sgrignoli as she left the terminal building, ran up to the window on Kempen's side of the car.

What are you doing? Where are you going? he shouted at the window.

Sgrignoli, horrified that she was about to be discovered, ignored him. The driver pulled away from the curb. Looking over at Sgrignoli, Kempen said, I could have him arrested. I should have had him arrested. Just this morning four photographers were harassing family members just trying to leave the hotel. These guys are unbelievable.

Sgrignoli could think of no response, and the two said nothing more to each other for the rest of the drive.

Once inside the hotel, Kempen showed Sgrignoli where the family briefings were and left her at the entrance to the ballroom. He had not showed her how to obtain the lapel button identifying her as legitimate participant in the family meetings, and there were some tense moments for her as a police officer questioned her and sent her to the TWA office to get the pin.

Credentialed, she returned to the ballroom where the medical examiner's representative, Bob Golden, was explaining what was being done to identify the bodies.

Sgrignoli was startled when Golden said he couldn't give a timeline for when the 100 bodies at the Suffolk County morgue might be identified and released. She was just as surprised at the level of agitation in the audience. People were shouting and interrupting the soft-spoken Golden.

"The briefing room was a real tinderbox that first day. And I was just dying because it was such an explosive story that the *Post* didn't have and that the families needed to have reported."

Moved by the emotions in the room and the uncertainty of her position there, Sgrignoli left the ballroom and went to the lobby bar for a drink.

For the first time since arriving, Sgrignoli was oblivious to the goings-on around her. Her thoughts were dominated by one question—do I stay or do I go?

Sgrignoli had come into the journalism profession relatively late, joining the *New York Post* in 1993 at the age of forty. Her passion for union activities had gotten her arrested while she was at the *New York Daily News,* and ultimately she was fired from her job as a copy editor. In the summer of 1996 Sgrignoli was still editing copy, rarely grabbing precious bylines.

Sitting at the bar, she was hit with the realization that she was in the middle of a career-making story. "Here I am, trapped on the copy desk in a situation I hate, and I get one day a week to report a story," Sgrignoli explained. "So when you're put on a big story that you know is going to be in the paper, it's like, 'Okay, now all I have to do is distinguish myself.'"

In a city chock-full of cutthroat reporters competing for scoops, it occurred to Sgrignoli that this scoop was hers.

"I should tell their story," Sgrignoli thought. "I want to tell their story and it will move people so that our coverage will make a difference." Her mind made up, she walked to a nearby pay phone and called her bosses at the *New York Post.*

Was she rationalizing, putting a noble spin on her own ambition? To this day the introspective reporter doesn't know. At the time it seemed like the right thing to do.

From where the families were milling about on the main floor of the Ramada Plaza Hotel, it was not possible to see the activity outside. When anyone left the hotel the contrast between the protected sanctity of the lobby and the fractious presence of a hundred reporters in the parking lot was exceptionally jarring.

By a breezeway on the side of the hotel, a row of cameras on tripods and lights on spiny stands were locked into position for the occasional foray by someone from inside the hotel. Reporters were desperate to match any new sliver of information from the nonstop news operations of CNN and MSNBC, a new all-news network that had gone on the air two days before the crash, so they furiously jotted down anything said by anyone who stepped in front of the cameras.

On Sunday night, following Dr. Wetli's unsuccessful presentation in the ballroom, a dozen family members stormed out to the press area blasting the medical examiner and the pace of the investigation. The media, which ordinarily use families for emotional dimension following a disaster, now found the tables turned. The families used the media as a bullhorn and it helped to diffuse the pressure inside the hotel.

At the East Moriches Coast Guard Station, another media camp had been set up in the gravel parking area outside the entrance. It had the look of a high-tech, low-rent trailer park.

Chubby satellite trucks formed a conga line of uplinks through the center of the lot, giving broadcast journalists the ability to "go live" at any time. Squeezed tightly together on either side were mobile homes and campers the spendthrift news outlets rented to use as portable offices. Publications with less money opted for the more rustic lawn chairs and card tables placed into empty U-Haul container trucks.

The necessities of any media stakeout—portable toilets and telephones—were in place twelve hours after the crash. From that point on, this sandy, dusty settlement was known as the litter box.

Photographers were lounging in lawn chairs or playing pickup football on the access road. They looked relaxed, but it was a ruse. Any official approaching the litter box would send them running with their cameras in a thundering dash, followed closely behind by the reporters, and the next round of media chaos would begin.

News executives don't like to admit it, but patience and circumspection are not highly valued on breaking news events. In the competitive frenzy following Flight 800's destruction, every minor development became a major story. This is especially clear in reviewing news articles from the summer of 1996. Anonymous criminal investigators were offering tantalizing bits of prophesy. "Prime Evidence Found That Device Exploded in Cabin of Flight 800" read one headline; "Plane Split in Sky, Officials Say, Suggesting Bomb," read another. The safety board's consistent warning that mechanical and criminal theories were being given equal weight was not only frustrating to ambitious reporters and their editors, it was boring. Substance was the loser in the ongoing battle with sensation. Nowhere was this more obvious than with the persistent speculation that a missile brought down the plane.

Captain Richard Russell, sixty-six, a retired United Airlines 747 pilot, never believed for a minute that the crash of Flight 800 was the result of some mysterious malfunction. He'd flown the 747 long enough to have an unshakable faith in its dependability. No, something more sinister had to be involved. Through friends familiar with the ongoing investigation, he came to his own conclusions. On August 22, he composed an e-mail explaining that the airliner had been hit by a navy missile. When he was done, he addressed the message to acquaintances in the air-safety community and hit "send." With a keystroke, Captain Russell launched himself into the investigation and the missile theory into a zone of credibility.

Russell never expected that his correspondence would impact the investigation the way it did. He was as surprised as anyone else when he learned it had caught the attention of former presidential press secretary Pierre Salinger.

The NTSB's Frank Hilldrup and Bob Swaim, who would be responsible for examining the airplane's systems, spent the days immediately following the crash ten miles out at sea, taking turns on the New York and Suffolk County police boats. They were towing sonar scanning devices that fed dense one-dimensional images to a monitor on deck. Hilldrup had enough experience in underwater recovery to be able to interpret the pictures and get a good idea of the wreckage lying beneath the water.

The *Pirouette,* a 110-foot search vessel, was operating twenty-four hours a day towing a six-foot-long torpedo-shaped pinger locator across miles of ocean. This one was more sophisticated than the hand-held one Deepak Joshi had used on the day following the crash, when he'd first gone looking for the black boxes.

Joshi had packed a few clothes and taken a bunk on the flat-bottomed work boat on Friday night. The accommodations were comfortable but far from plush. The four meals a day prepared by the boat's cook were said to be so good that navy officers soon arranged for planning meetings to be held at dinnertime on the *Pirouette.*

On Monday, the pinger locator torpedo was retired. No returns had been detected. The experts concluded the noise-emitting devices on the black boxes were either buried under wreckage and inaudible, or inoperable from impact.

The new plan involved more dragging of sonar scanning devices and the use of a large navy salvage ship, the *U.S.S. Grasp,* which was to be ready for operation the next day. Laser scanners that create much more literal images were added to the technology in use. Videotapes created from the laser and sonar scans would be used to make maps of the debris fields, to identify pieces for retrieval, and to locate bodies.

The natural impulse following a disaster like this one would be to jump in and start pulling up wreckage, but experts with the navy and the NTSB insist that preparing for wreckage retrieval is crucial to effective use of divers, boats, and equipment. Analysis of the wreckage field allows investigators to determine what parts of the plane came off first and, sometimes, even how.

Though the tremendous effort going on in the Atlantic should have provided a sense of purpose and progress, it also heightened frustration. The flight and cockpit recorders, for example, could not be found even though the Air National Guard pilots had noted the exact GPS (global positioning system) location of the tail sinking into the Atlantic.

On the 747, as on most every other type of jetliner, the FDR and CVR are located in the rear of the plane. The location noted by the guardsmen should have focused the search for the boxes, but neither the navy nor NTSB officials could explain why this information was not used.

At the Ramada Plaza Hotel, family members were also wondering why so much activity was showing so little results. They packed themselves into the ballroom three times a day for presentations, delivered in halting spurts of English, French, and Italian. On one of the first nights, Francis arrived at the hotel and launched into a detailed recitation of the recovery efforts, only to be repeatedly interrupted by hecklers.

Retrieving bodies from the Atlantic went on throughout the night. Four Coast Guardsmen on the boat on the right almost became victims themselves. Aircraft wreckage ensnared the propellers of the twin engine Monarch, and the disabled boat was nearly swept into the flames.

Captain Christian Baur was flying in the left seat of the 106th Air Rescue Wing Pave Hawk helicopter when he saw a pin of light erupt into a wall of flames over the southern shore of Long Island. Debris was still falling through the sky when the helicopter reached the scene.

Early on, bodies were recovered quickly. One hundred were taken to the Suffolk County Medical Examiner's Office the morning following the crash. Passengers who were submerged with the airplane took longer to find.

Dr. Charles Wetli explaining to the families of Flight 800's victims why it takes so long to identify and release the bodies. His boss at the time, Suffolk County Health Department director Dr. Mary Hibberd, stands behind him.

NTSB Vice Chairman Robert Francis worked tirelessly as the agency's spokesman at the crash scene, conducting two and three news conferences a day in addition to meeting with victims' families. Back in Washington, however, Francis was criticized for making decisions that should have been left to the investigators.

Navy divers Kevin Oelhafen (*left*) and Douglas Irish (*right*) help Brad Fleming prepare to dive off the U.S.S. *Grasp* on July 25, 1996. Oelhafen and Irish discovered the 747's black boxes the night before. (*Photo by Airman Charles Withrow, USN*)

Bernie Loeb (*left*) and Vern Ellingstad of the NTSB show the battered cockpit voice recorder and flight data recorder from Flight 800 to reporters gathered at agency headquarters in Washington, D.C.

Recovering bodies and aircraft wreckage was gruesome and dangerous work for divers. One said it was like walking through razor blades. Few were prepared for the sight of the passengers' remains. Remarkably, through four months of diving, there were no serious injuries. (*Photo by Glen J. Hurd, USN*)

The U.S. Navy recovered 95 percent of the wreckage from the Atlantic. Workers from the NTSB, FBI, Boeing, and the pilots', flight attendants', and machinists' unions spent months putting it all together. (*Photo courtesy of NTSB*)

The FBI's Jim Kallstrom ends a briefing with reporters who camped at a gravel parking lot outside the Coast Guard station in East Moriches. Thousands of journalists descended on the resort communities known as the Hamptons in the days following the crash.

From the outside, the 747 is a symbol of modern technology, an icon of the jet age. But the cockpit of a Boeing 747–100 reflects its three-decades-old design, with analog technology and electromechanical displays. The 747 is one of the oldest commercial Boeing designs still in production.
(*Photo courtesy of Boeing*)

Captain Steven Snyder, shown here in 1985, became a pilot with TWA in 1965. He was so interested in fuel economy, pilots often referred to flying techniques used to minimize fuel consumption as "Snyderizing the plane."

Ralph Kevorkian of Garden Grove, California, was about to become certified as a 747 captain after years in the left seat of the L-1011. Flying the 747 meant more money and more prestige. It was an obvious move because TWA had announced plans to eliminate the Lockheed wide-body from its fleet.

Oliver Krick, seated at the 747 flight engineer's desk, and instructor Richard Campbell had flown several flights together. This photograph was taken by Krick's mother, Margaret, as Krick and Campbell prepared for a flight to Tel Aviv the week before they died.

Ray Lang was one of just a few male flight attendants when he joined TWA in the mid-seventies. He enjoyed the travel and meeting new people. On Flight 800, Lang was working with Melinda Torche. The two had plans to marry.

Susan Hill, shown here in 1985, was a homicide detective in the Portland, Oregon, Police Bureau. Her trip to Paris was her first vacation following her divorce. She tried to convince her mother and a girlfriend to accompany her but wound up traveling alone on Flight 800.

Fashion designer Monica Omiccioli and banker Mirco Buttaroni were married on June 23, 1996. Their three-week honeymoon to the United States was the first flight for both of them.

In the fall of 1995, Charles Henry "Hank" Gray, 47, was on vacation in San Francisco with his girlfriend, Tara Tomlin. He often predicted he would die before his fiftieth birthday.

Judith Yee and her dog, Max, in New York. Yee handled the travel arrangements that put her, her cousin Patricia Loo, and friend Angela Murta on Flight 800. She chose TWA because it allowed caged pets to travel in the passenger cabin. (*Photo by Paul Coughlin*)

Daniel Cremades, a Spanish citizen living in France, started taking college courses at age 14 at Cal Tech in 1995. He was returning home from a second summer of studies at the Massachusetts Institute of Technology.

Monica Omiccioli's uncle and business partner, Jean-Claude Poderini, speaks to reporters in New York. He was one of many angry family members who blasted the Suffolk County Medical Examiner for taking too long to identify and release the bodies of Flight 800's passengers.

Students at Montoursville High School in Pennsylvania grieve at the news that sixteen members of the school's French Club and five chaperones were killed on Flight 800.

By the end of the summer of 1996, National Transportation Safety Board chairman Jim Hall came to believe the mystery of Flight 800 was the NTSB's to solve. "We were in it for the long haul," he acknowledged. *(Photo courtesy NTSB)*

TWA's Special Health Services director Johanna O'Flaherty was roundly praised for the work she did helping the families in the weeks following the crash. But the airline's initial insensitivity and the delay in releasing an official passenger list prompted Congress to enact legislation on behalf of the victims of airline disasters. *(Photo by Kathleen Svoboda)*

Airport workers at the special operations center of Charles de Gaulle Airport in Paris handle calls. At the Paris airport, a detailed disaster plan had been created following the bombing of Pan Am 103, and as a consequence TWA was far more prepared in Paris than in the United States.

They're smiling in the seventies, but John Borger, Pan Am vice president for engineering (*left*) with his boss, Juan Trippe, in front of a new Boeing 747–200, took a huge financial gamble in 1966, investing in the development of the jumbo jet.

Boeing's 747 being assembled in Everett, Washington.

Pan Am Captain Eugene Banning was presented with an air safety award from the Air Line Pilots Association in 1971. After the crash of Pan Am 214 in 1963, Banning tried to convince aircraft manufacturers and the FAA that fuel tank inerting was essential for safety.
(*Photo courtesy of Air Line Pilots Association*)

Flames from a Pan Am 707 that crashed in Elkton, Maryland, in 1963 after lightning touched off an explosion in a wing fuel tank. Air safety investigators suggested fuel tank inerting to prevent similar accidents, but the FAA did not adopt the recommendation.

The large metal structure with portholes is a mock 707 wing tank created by Parker Hannifin to demonstrate its liquid nitrogen inerting system. Aircraft manufacturers including Boeing, airline representatives, and the FAA attended exhibitions like this one in Los Angeles in 1969. Boeing was not persuaded to incorporate the system on the 747.

Attendees at the Parker Hannifin demonstration posed for a photo in the test hangar. Hannifin's Cleve Kimmel is on the far right. The FAA's Tom Horeff is tenth from the right in the back row. Boeing's Ivor Thomas, bearded, is kneeling in front.

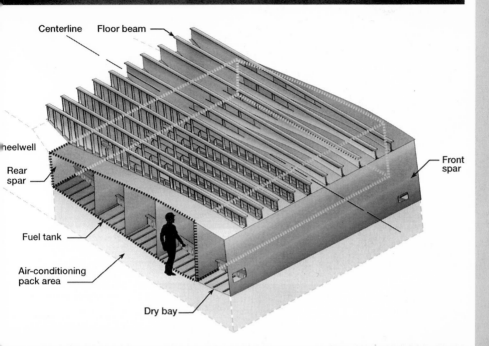

Centerline   Floor beam

heelwell

Rear
spar

Front
spar

Fuel tank

Air-conditioning
pack area

Dry bay

The center wing tank on a 747–100 holds 12,890 gallons of fuel. On flights of six hours or less, this tank is often flown empty. The figure in the top illustration indicates tank height, not access. Maintenance entrance is through small openings in the front wall. (*Illustrations courtesy of Boeing*)

Pierre Salinger made international headlines by announcing he had top secret documents proving a missile brought down TWA Flight 800. When questioned by CNN, however, Salinger admitted his document was the same as one that had been on the Internet for six weeks.

Captain Richard Russell, with one of the planes he restores as a hobby, helped launch the missile theory by sending an E-mail to a dozen friends. A retired 747 pilot, he is convinced nothing inherent in the airplane could have triggered the crash.

Major Frederick Meyer was the only one of nine Air National Guardsmen flying on July 17, 1996, who claimed to have seen military ordnance hit TWA Flight 800. The Vietnam veteran's observations became a crucial part of persistent theories that the U.S. government was covering up the true cause of the crash. (*Image courtesy of CNN*)

Elizabeth and James Sanders were tried and convicted for conspiring to steal Flight 800 wreckage. Jim Sanders claimed it was a First Amendment issue, but the jury disagreed. The couple was sentenced to probation on the eve of the third anniversary of the crash.

Ed Block, former wiring procurement officer with the Pentagon, is a crusader for tougher standards on aircraft wiring. Shortly after the crash of Flight 800, Block became convinced that flaws in the airplane's wiring were somehow responsible for the explosion in the fuel tank. *(Photo by William Johnson)*

When TWA Flight 800 exploded, NTSB fire and explosion expert Merritt Birky *(left)* was still investigating whether oxygen canisters caused the catastrophic fire aboard a ValuJet DC-9 in May.

TWA Captain Terrell Stacey was so concerned about the integrity of the crash investigation that he removed material from the plane wreckage and gave it to a journalist for private chemical testing. He pleaded guilty to federal charges and returned to TWA as a pilot in 1998.

The NTSB's Bob Swaim uses a magnifying lens to examine miles of wiring from N93119. His relentless pursuit of the source of the ignition of the fuel tank lasted more than three years. *(Photo courtesy of NTSB)*

Deepak Joshi, examining the structural elements of the plane, stands inside the center wing tank, which has been partially reconstructed.

NTSB investigators spent weeks at a time away from home working at this hangar in Long Island. From left, Bob Swaim, Ron Schleede, Deepak Joshi, Merritt Birky, and Jim Wildey.

Bill Brookley was severely burned in an Air Force plane crash in 1956. Three others were killed. A decade later he began working on the development of explosion and fire suppression systems on military aircraft, including the F–15 Eagle, a model of which he is holding. Brookley believed fuel tank inerting would have saved the lives of his fellow airmen. (*Photo by Dione Negroni-Hendrick*)

Ron Schleede (*left*) and Al Dickinson of the NTSB at the public hearing for a United Airlines crash in Colorado Springs in 1991. Dickinson was the safety board's investigator in charge on the Flight 800 investigation, but others were making decisions for him.

Robert Clodfelter described his work in fire prevention for the Air Force as lifesaving. By the eighties, he thought what the Air Force knew about explosion suppression on combat planes was ready to be applied to commercial airliners, but the FAA had no interest.

Thomas McSweeny, the FAA's director of aircraft certification, testified at the NTSB hearing on Flight 800 in December 1997, suggesting the FAA would reconsider its earlier criteria for fuel tank safety.

Boeing's position at the time of the hearing was that the risk of a fuel tank igniting from an electrical malfunction was extremely low. Engineer Rich Breuhaus was assigned to review the plane's systems and see what hazards might have been overlooked. *(Image courtesy CNN)*

Ivor Thomas had been a fuel systems engineer with Boeing for thirty-one years when he testified for the company at the NTSB public hearing in December 1997. Days later he offered his services to the FAA, helping regulators redefine acceptable risk for fuel tank flammability.

Passenger seats from the TWA 747 were placed in position inside ninety feet of reconstructed fuselage as an investigative tool. It created a powerful display of the kinds of forces to which passengers were exposed.

NTSB chairman Jim Hall watches as the ninety-foot reconstruction of N93119 is moved to a smaller warehouse for storage in September 1999. So painstaking was the reassembly work and so dramatic was the result that Hall plans to make it a centerpiece for a new NTSB aviation safety center.

"We get lots of information, but we don't want the information they give us," Monica Omiccioli's uncle Jean-Claude Poderini said. "We want the bodies, and we want to go home."

Joe Lychner explained how family members' hopes hung on these meetings. "You didn't want to miss one of those briefings because you were holding on to threads of information."

Standing in the packed ballroom, they listened to the details of dives accomplished and wreckage located. But the words were just drops of water in an ocean of grief.

The media and the public were obsessed with whether the disaster was a crime or an accident. Yet Lychner explained that the constant focus on the cause made the families feel even more isolated.

"Our number one goal was to get the bodies back. Sure, we wanted to know what happened, and we wanted to bring the people to justice who were responsible for it. But we were terribly afraid the FBI was spending too much time on the investigation and that was taking priority away from the recovery of the bodies. In the ocean we knew our loved ones could just float away."

This concentration of attention is not unexpected, according to experts in the area of trauma response.

"It is common to fixate on the recovery of the bodies. Finding the remains is important to people because it provides finality, even if it's a painful finality, it's a marker," explains Dr. Mary Hibberd, Suffolk County's health director at the time.

Having been moved by Lychner's loss, Eileen Long of the governor's office arranged for him to be picked up at the airport on his arrival from Houston and taken to the Ramada. She asked Marlin to find him when he arrived. Like Long, Marlin found he was touched by Lychner's story and impressed by his sense of proportion and natural leadership.

In the days that passed, Marlin, Governor Pataki, and his staff began to rely on Lychner's impressions of what was going on with the families at the hotel. And Lychner was relentless in taking advantage of their attention, explaining time and again that recovery of the bodies was the number one priority.

It was a message Pataki may have wished he did not hear so well, because in trying to address those concerns, he caused the families a lot of grief. Though he turned out to be right, the episode arguably hurt his credibility. This happened as he was bringing what he considered good news. He told reporters, after a large beachside memorial service on Monday, that more bodies had been spotted by divers near a large piece of wreckage.

"A large number of bodies are trapped under or within the debris and they will be brought to the surface as soon as that is possible."

The journalists, who'd been told just the opposite, were incredulous, one asking, "I don't mean any disrespect, Governor, but are you sure?"

Pataki did not budge.

"I was told this by Mr. Kallstrom," the governor answered.

It is exactly the way Jim Kallstrom remembers it.

"We were at the memorial and I got beeped," Kallstrom explained. "Before Pataki speaks, I find out that the navy found the major debris field, which was big news. We thought we'd probably find the bodies in the debris field."

Two hours later, Pataki expounded for another set of reporters at the East Moriches Coast Guard Station.

"We hope sooner rather than later that the other remains that have been identified, and there are dozens and dozens, and it could be, I don't want to be overly optimistic or raise false hopes, but it could be as many as one hundred additional bodies down there, that they will be able to recover and bring to the surface."

At the Ramada Hotel that afternoon, the governor's more detailed announcement was greeted with a glee comprehensible only among the bereaved. One escort who'd befriended several families was at a loss at the bizarre contrast of emotions when one woman learned her loved one had been identified.

"She came running up to me and she had a smile and said, 'This is wonderful they found my brother!' 'Why would that be wonderful news?' I'm thinking. In hindsight I recognize that all sorts of horrible thoughts were going through her mind. And then I understand why knowing his body was found had made her so happy."

Whatever joy and expectation Pataki's comments triggered, the relief was fleeting. Bob Francis was furious, denying there were any bodies seen that had not been recovered.

"That was the worst night of the investigation," Francis said. "I called him up and told him that what he said was less than helpful."

It wasn't so much what Pataki said that angered the safety board, it was what his comments implied.

The NTSB's temporary press office was operating out of a first-floor meeting room behind a curtained sliding-glass door just off the lobby of the Sheraton Hotel in Smithtown, a fifty-minute drive from East Moriches. Shelly Hazle, at twenty-five the youngest and newest staffer in the public information office, said she struggled to correct the impression Pataki's comments made while remaining deferential to the sensibilities of the victims' families.

"I was getting a lot of calls from people who thought the plane was in big pieces. They thought the passengers were all strapped in their seats, everyone just trapped inside. Naturally they wanted to know, 'Why aren't you bringing them up?' But that's not the way things were."

Pataki's description that the bodies were trapped in wreckage was based on more than what Kallstrom had whispered to him during the memorial service. New York State Police divers had seen the victims. One was quoted in that morning's *New York Times* describing them as "not recoverable—some are pinned inside—until several large pieces of wreckage are moved."

"We are confident that there are a lot of bodies down there, but we just don't know how many right now," Kallstrom told a reporter, adding, "This is exceptionally good news for the victims' families."

The governor's staff maintains that one reason Francis denied the news was that the safety board did not want to have to explain that the bodies which had been underwater, decomposing and exposed to marine life, were going to be difficult to retrieve. To complete the gruesome and time-consuming task, large pieces of the aircraft were going to have to be lifted first.

So many politicians were on television during this time, the episode with Pataki was inevitable. At the White House, Kitty Higgins was worried about the lack of control over what was being said publicly.

"It was chaotic because you had the state of New York, the city, the various county people and law enforcement and all the politicians who wanted their faces out there. It was a feeding frenzy."

The stir created by Pataki's announcement was over in a day or two. The comments from Congressman Michael Forbes of New York, however, had far-reaching consequences. Forbes drove all night from Washington to be back in his district the morning following the crash. He hadn't been on the scene for long when he announced, incorrectly and on live television, that one of the black boxes had been found. The safety board spokesman, Bob Francis, denied it.

Asked that night whether he'd been wrong about the discovery of the box, Forbes would not back down. "It's been located. I don't know if it's actually been pulled up at this point or not. That we have not been able to confirm."

Forbes's administrative assistant, Kelly O'Meara, says an angry Forbes ordered her to find out why he had been told one thing and the press another. She was reluctant at first, but after months of digging, requesting information on behalf of the congressman from a number of agencies, she came to the conclusion that the box had been found but taken away for secret analysis.

The order from Forbes to investigate the investigators started O'Meara on an assignment that lasted more than a year and led her to believe there was a cover-up of the true cause of the crash. She shared her suspicions with reporters. As a consequence, many of the news reports suggesting a missile caused the TWA disaster had their origin in the office of Congressman Forbes. O'Meara had a position of credibility and substantial access to information. Her allegations had as significant an impact on the investigation as those of Pierre Salinger.

Officials sought to calm families caught in the crossfire of conflicting information with a promise they would be briefed on the progress of the investigation before the press. It was a quixotic proposal made by the very people leaking information to select reporters. A large dose of distrust was added to the insistent sorrow at the hotel.

A week had passed without recovering the 747's black boxes, and progress was only slightly better on two other priorities: recovering the remains of passengers and getting a picture of how the wreckage of the plane was laid out on the ocean floor.

When Kevin Oelhafen stepped off the diving platform lowered from the *Grasp* and saw through an opaque iron-colored ocean an orange box with FLIGHT RECORDER stenciled on the side, his first thought was that things were looking up. Topside, Red Diver, he identified himself through a microphone in his diving helmet, I found the flight data recorder.

Oelhafen stood near the FDR and directed the operator on the *U.S.S. Grasp* to move the remote vehicle and camera to what he'd just found.

On the fantail of the *Grasp,* the NTSB's Frank Hilldrup heard a shout from one level below where sailors were viewing the camera transmission. They've got one of the boxes! he heard someone say. Dashing down the stairs, he saw Oelhafen on the television, pointing to a box on the floor of the ocean. Ask him to hold it up, he said to the sailor on the radio, thinking the scene could not have been sweeter. As Oelhafen began to lift the box, his diving partner, Doug Irish, came into view. In his arms he carried another orange box.

No "atta boys," no "way to go's" went down to Oelhafen and Irish. But up top Hilldrup was unrestrained in his glee. "Fucking A, they've got both of 'em!"

That comment is preserved on the videotape of the event, and had he known that, Hilldrup might have chosen his words more carefully. He explained, "I was ecstatic just like everyone else." The two divers, isolated from the elation on board the *Grasp,* were pleased that the navy had accomplished something important for the investigation so soon after arriving. In their lead-lined boots, each toting a heavy air hose, the men made their way to the lift and tied the recorders to the basket. When they were done, they still had time on the clock. This dive was intended to recover bodies. Though the FDR and CVR were gratifying finds, they could see passenger remains that needed to be brought up, too.

Dennis Grossi is the man who reviews flight data recorders and turns the information into a language investigators can understand. A genial, middle-aged man, he is not bothered in the least that he labors in a tedious and obscure specialty. He takes the occasional high-profile crash tapes when they come in and doesn't complain about the overwhelming number of boring ones. To Grossi, turning a thick stew of zeros and ones into a clear picture of a plane's last thirty minutes is always rewarding.

Both the flight data and the crew conversations on N93119 were recorded on magnetic tape, surprisingly similar to the type used on old-fashioned reel-to-reel recorders. The flight information is analog, but in labs like Grossi's at NTSB headquarters it is converted to digital as it is read back. Everything the plane did, from its air speed to its rate of vertical climb, is there, just waiting for someone like Grossi with the vision and the patience to figure it out.

Jim Cash's job, assessing the voice recorder, seems easier. Beyond the simple task of listening to and transcribing the conversation on the flight deck, however, is the more complicated interpretation of the noises that were recorded at the same time. The tape of Flight 800 was both straightforward and baffling.

Cash explained that in-flight breakups usually give some warning to the crew. "The most intriguing part was the lack of anything intriguing," he said. "Usually, something is going on before. The crew will say something or there's other indications going on. In this one, everything was just perfectly normal, just the noise at the end and that's it," referring to a split-second sound at the end of the tape.

Cash and his group couldn't get very far with the noise. "We reviewed it using ear, meter, spectrum charts and graphs." In the months to follow they would compare the noise against sounds recorded on other aircraft that had exploded. But Cash wasn't optimistic that his group would give much insight into the cause of the crash.

"About the third day I had the recorder, it was obvious that I didn't have the golden nugget and probably wasn't going to have the golden nugget," he concluded.

14

Air travel, in the competitive, cost-cutting, post-deregulation era, is far removed from the heady, exciting first years of overseas jet travel. In the early sixties, passengers dressed elegantly, dined on steak, and were pampered by attractive, solicitous stewardesses. A new word, jet setter, was coined to describe this elite air traveler. The allure of this glamorous mode of transportation was obvious. Passenger volume increased fifteen to twenty percent a year.

Then came the 747, a creation to redefine high-capacity flights and improve airline economics. John Borger, Pan American World Airways' vice president for engineering at the time, describes what airlines wanted from the 747.

"The big objective was to reduce operating costs. We wanted to operate through the same airports as the 707 and make the cost per seat mile thirty-seven percent lower than the 707."

Attempting to meet that objective was a phenomenally daring move for the Boeing Company, and the biggest financial risk it had ever taken. Failure would have doomed the company.

The plan for a super-sized passenger aircraft was, in part, a way for Boeing to cut its losses after failing to win an air force contract to build a big troop carrier designated the C-5A. When Lockheed got the deal, Boeing was left with expensive design work for a plane it would not build, unless, of course, it turned the military project commercial. The business forecast provided incentive. In addition to booming passenger volume, the demand for air cargo service was growing even faster.

Two aviation visionaries were behind the decision to build the 747: Boeing's chief executive, William Allen, who bet $2 billion on the plane, and Pan Am president Juan Trippe, who joined Allen in the gamble with an initial order that provided Boeing with enough cash to get the jumbo jet going. For Boeing to get back its investment, it would have to sell at least 150 747s, one quarter of the number of 707s it had sold during ten years.

Trippe was notorious for his ability to manipulate business deals to his advantage. From Boeing he purchased the right to be the first American airline with the 747 and a large say in the plane's design. From the day in December 1965 when Trippe and Allen shook hands on an agreement, Pan Am designers and engineers became part of the team building the 747.

Trippe had plenty of reasons to want his airline involved in every phase of the progress, all of them green. His company was at risk, too. Pan Am's first order for twenty-five 747s at a cost of $500 million was more than it had made in its most profitable year.

Thirty years to the day before the crash of Flight 800, at the newly constructed Boeing assembly plant in Everett, Washington, Boeing hosted an open house for the press and public. The main attraction was a completely furnished and decorated plywood mockup of a 747. Describing the event, Laurence Kuter, a Pan Am vice president and retired air force general, reflected with military reserve on the optimism of the day.

"For the first time the wide-body concept with its double aisles and spaciousness was exposed to public comment. The comment was gratifying."

That day was also notable for the rare display of unity between Pan Am and Boeing. During the three-year project, disputes far outnumbered agreements. One of the biggest problems was the weight of the 747. The plane had grown heavier as Pan Am added features to the original design, until the 747 was so heavy it could no longer perform as promised.

"Weight is the biggest penalty an airline can carry," Boeing spokesman Gary Lesser explained. "The tax you pay for having additional weight is reduced revenue potential."

The 747 was built around new high-bypass Pratt and Whitney fan engines that also pushed existing technology until, according to retired engineer Steve Hatch, the 747 pushed too far.

"We had to have an advanced engine in the beginning, thrust-wise, because the airplane got too heavy."

The plane's upward spiraling heft would have to be reversed.

Pan Am was making demands on both ends. It desperately wanted its first 747s delivered in time for Christmas 1969, to take advantage of seasonal swells in passenger traffic. At the same time, it was unwilling to back off on issues that affected the plane's weight. Pan Am threatened to withhold payments. Boeing responded it would go bankrupt.

One year after the wooden mockup of the 747 was greeted with awe, Pan Am and Boeing were preparing to sue each other.

Captain Eugene Banning was one of the Pan Am employees sent to Boeing to help design the 747. The middle-aged 707 pilot spent the next two years crisscrossing the country from his home in New England to Boeing's factory in Everett, Washington.

Pilots were brought in for advice on the cockpit, to be consultants on the location of instruments, the operation of controls, and the lighting on the flight deck. Pilots from other airlines purchasing the 747 also contributed, including pilots from TWA. Banning's interest went beyond the cockpit.

Fewer than three years had passed since he'd stood over the wreckage of Pan Am Flight 214 and made a silent commitment to do something to prevent fuel tank explosions. He'd taken a seat on an FAA committee studying the feasibility of fuel tank inerting and, at the same time, convinced his union to make an official call for explosion prevention devices on all new airliners. Two years later, the union demand was specifically addressed to Boeing and the 747.

"We made a request for an inerting system, initially with the hope that one could be found," Banning explained. What he did not know was that Boeing had already seen the most promising fuel tank inerting technology available and dismissed it.

Eleven hundred miles down the Pacific coastline from Seattle, the Parker Hannifin Company had been energetically marketing a system to address the problems raised by the crash of Pan Am Flight 214.

"Because the 747 and the 707 have center tanks, they are more dangerous than most military airplanes because you are sitting on thousands of gallons of fuel," explains Hungarian-born fuel systems engineer Attila Lothrigel, who helped create a process of deoxygenating fuel tanks to make them explosion-proof.

The system displaces oxygen in the fuel by pumping nitrogen through it. The oxygen enters the vapor space above the fuel, called the ullage, and it escapes through the vents. No oxygen, no explosion. That's the way it works.

Using liquid nitrogen this way was not a new idea. Parker Hannifin's version was originally designed by North American Aviation Company* to protect the tanks on the XB-70, a fifties-era high-altitude bomber developed for the air force. The plane

---

*North American Aviation became North American Rockwell in 1967. The company is now called Rockwell International.

flew at three times the speed of sound, generating enough friction over the wings to pose a risk of heat igniting the fuel tanks. Explosion-proof tanks were crucial.

The success of the inerting effort was dramatically and tragically illustrated on June 8, 1966. A prototype XB-70 was being flown with four other aircraft for a publicity photo when an F-104 flew into its wing. Both planes were lost along with their crews. The F-104 exploded, but there was no fuel explosion or fire on the crippled XB-70, even though all six engines were running.

The air force soon lost interest in the XB-70 and the program was killed. Parker Hannifin, confident in the potential of fuel tank inerting, hired the scientists who developed it, with plans to market a system to the commercial airline industry. The hazards of fuel tank explosions were known. The debate had been over how much protection was sufficient.

Boeing helped Parker Hannifin by giving the company the kind of design and fuel system information it needed to adapt the system to a 707. Ted Tsue was one of the Boeing engineers who spent time analyzing the system and helping the Parker scientists prepare their presentations.

"My interest was to understand what's out there to protect my fuel system," he said.

Because of their own experience with fuel tank explosions, Pan Am and TWA representatives were also interested. They were the first to see the finished prototype.

Jim Lowes, the vice president of marketing at Parker Hannifin, remembers a meeting with Juan Trippe at his office in the Pan Am building in New York. It was shortly after the crash of Flight 214.

"The president of Pan Am, Juan Trippe, would say, 'We want to put this on. We think it's the right thing to do. If we had it before, we wouldn't have lost Flight 214.'"

The 707 wing tank fuel system sat behind Parker's manufacturing plant. It was an unsightly but functionally accurate display that showed how the fuel was scrubbed as it was loaded into the tank. Spark plugs would be used to try and ignite the space above the fuel, called the ullage, while viewers looked through Plexiglas portholes. It was perfectly safe, because the vapors would not catch fire.

TWA's David Hartline, senior project engineer for fuels and lubricants at the time of the Rome crash, saw the show and went back to Kansas City with the feeling that the system was too complex. At the same time, he admits, it was hard for him to see beyond the specific tragedy his airline had experienced. The devastating fire aboard the 707 had been caused by flames rushing into the center tank through the wing vent.

"I felt we were solving the specialized type of explosion we experienced, and so we localized on fixing that very problem."

Boeing engineer Ivor Thomas also visited Parker Hannifin. Like Hartline, Thomas thought nitrogen inerting was overbroad. As far as he was concerned, the hazards exposed in the TWA and Pan Am accidents could be addressed by fixes in the vent lines.

"Most history was from fire coming into the vent system from outside, like the Pan Am lightning strike and the TWA flight in Rome," Thomas said, explaining TWA's choice of a simpler fix, the installation of a Fenwal flame-suppressing system in the wing vents. With this, any ignition entering the fuel system through the vents is sensed and extinguished.

When the 747 was designed, the vent outlet was located farther in on the wing, and the FAA determined the risk of fire or lightning entering the plane through the wing vents was eliminated. Even so, TWA was so committed to the Fenwal system that it paid extra for Boeing to install it on the nineteen 747s it bought, including the plane that flew as TWA Flight 800 on July 17, 1996. The center wing tank on the flight to Paris was not ignited through the wing vent, of course, so the Fenwal suppressors never triggered.

"We applied a localized fix for our problem without meeting the overall explosive problem of the air space above the fuel," explained Hartline. "We were ostriches with our heads in the sand solving the problem that concerned us because it had happened to us."

Ivor Thomas was a whiz-kid who looked even younger than his twenty-five years when he arrived at Boeing in 1966. He'd started college at sixteen, and had been working for several years with the Bristol Aircraft Company developing the Concorde. Ready to try his own wings, he moved to Seattle and took a job as a fuel systems engineer on the 727 and later the 707. When it came to Parker Hannifin fuel tank inerting ideas, Thomas's experience in fuel systems was balanced with a practical understanding of airline economics.

"Technically, it was a feasible system, but did you need it? Was the cost and the weight worth it?" Thomas wondered. "Those decisions got mixed up with carrying the heavy system around."

The machinery and tanks of nitrogen sufficient to inert the 47,000-gallon capacity of a 747 for one coast-to-coast flight would add about 1,450 pounds to the plane's takeoff weight.

And then there was the cost. Nailing it down was not easy. Parker Hannifin estimated it at $15 to $20 for one 747 flying one trip, coast-to-coast. The airlines thought it would cost much more, considering operational and maintenance costs, as well as the fact that the additional weight displaced an equal amount of payload.

By 1967, the engineers and salesmen at Parker Hannifin were convinced the FAA was soon going to require inerting on new model aircraft. The regulators had already solicited Parker for information to help write the rule. As far as Parker Hannifin executives were concerned, the threat of a federal requirement should have been enough incentive for Boeing to buy their product. Lowes and Parker Hannifin engineers Attila Lothreigal and Cleve Kimmel traveled to Everett to meet with Boeing, bringing a seventy-five-page pitch with them.

"We made an unsolicited proposal. We said, 'You ought to have this system on the airplane as it's coming off the line and not do it on a retrofit basis,'" Lowes says his men told the Boeing engineers. "Do it now," was the suggestion, "you'll have to do it eventually."

Getting an inerting system on board the new 747 was an all-consuming project for Lowes and the others. At Boeing, it was just another add-on and a heavy one at that. At Boeing, the pre-occupation was on saving weight. Steve Hatch, who worked on the 747's original design, says engineers were told they could spend money to save weight.

"Two hundred dollars a pound for one hundred pounds" was the formula, he says. "You could spend two hundred thousand dollars on the changes. You could go buy something or put something on that causes our costs to go up by two hundred thousand dollars to save one hundred pounds."

Creating the 747 was a complex undertaking already. Boeing employees were just hoping to get the plane designed, built, and flying before it put them all out of business. Ultimately, neither Boeing nor Pan Am considered a fuel tank explosion suppression system important enough to slow down progress on the 747.

By the end of the sixties, Hannifin sales people realized that short of an FAA order, no airline or airplane manufacturer would go out on its own and invest in fuel tank inerting, not for new model airplanes and certainly not for planes already flying.

"The airlines were not convinced that they had a safety problem and to ask for a proposal for an inerting system would be tantamount to admitting that their aircraft was unsafe," explains Parker Hannifin's Kimmel, now retired.

Airlines might have been unwilling to see the problem, but the Air Force couldn't miss it. In 1967, when large numbers of aircraft in Vietnam were being lost in what should have been nonlethal attacks, the Air Force Aeropropulsion Laboratory at Wright Patterson in Dayton, Ohio, was assigned to find ways to protect fuel tanks on the planes and helicopters being used in the conflict. Solving the problem was given the highest priority.

"The military worries about the explosive nature of fuel tanks since the planes are often flying in hostile areas. Even relatively minor gunfire could penetrate a fuel tank and cause a disastrous explosion," explains Robert Clodfelter, a civilian scientist working in research and development at the base at the time.

"When you count up the loss of life, aircraft, and missions, the number was something like two thousand," remembers Clodfelter. "We were losing all these planes to small-arms fire."

The labs tested a hodgepodge of methods, from the unconventional—whiffle balls—to the unsophisticated—chunks of polyurethane. All kinds of materials were stuffed into the fuel tanks to quench flames and prevent the buildup of pressure that creates a destructive explosion.

Since planes with large tanks posed special problems, Parker's liquid-nitrogen-inerting system was selected for the C-141, a Lockheed transport that had just entered Air Force service, and the KC-135, which shared the fuel system design of the 707.

The job of adapting the system to work on Air Force planes was given to a man who months earlier had joined the Aeronautical Systems Division's fire protection and suppression branch at Wright Patterson as a civilian engineer.

The assignment could not have been more appropriate for William Brookley. He had been a flyer himself, a flight engineer on an air force troop carrier. And he'd been severely burned in 1956 when the C-124 Globemaster in which he was flying crashed and caught fire. He spent nine months in the hospital recovering, but three of the men flying with him died.

"If there had been a fuel suppression system on the plane, all of us would have walked away," Brookley recalls. "We could have climbed out of the aircraft."

15

Of the eight men flying Air National Guard aircraft with the 106th Rescue Wing on the night of the crash of TWA Flight 800, only two had participated in actual wartime rescues. Dennis Richardson, fifty, and Frederick "Fritz" Meyer, fifty-six, had both served in Vietnam. An air force sergeant, Richardson had received three Distinguished Flying Crosses and three Air Medals. He also won a Purple Heart trying unsuccessfully to rescue a fighter pilot downed in North Vietnam in 1967. His HH-3E twin-engine helicopter had been chased off under heavy gunfire while attempting the rescue, and Richardson was shot in the arm.

After the crash of Flight 800, Richardson reflected on the difference between what he did as a soldier in the 1960s and the things he saw that night. "In 'Nam, it's a war. If you're going to fly a rescue helicopter in Vietnam, you know some of us aren't going to make it back. But that night, there were so many bodies in the water, kids and some with not all of their arms and legs, there was nothing we could do for any of them. There was no one left alive."

Not far from where Richardson was putting his life on the line in 1967, Fritz Meyer, a navy lieutenant, was piloting similar daredevil search-and-rescue missions in North Vietnam. His unit was supporting a navy helicopter assault squadron along the Mekong River. "I remember my first mission in the area, it was flying body bags out. It was a horror story," Meyer recalled. "I made plenty of pickups in hostile territory, but no one who ever flew with me was ever injured, and my aircraft was never damaged from gunfire," he reported proudly years later.

Like Richardson, Meyer earned military decorations: a Distinguished Flying Cross, six Air Medals, and a commendation from the Secretary of the Navy. The experience and prowess displayed by Meyer and Richardson during the conflict in Southeast Asia served the National Guard's 106th Air Rescue Wing well.

"That's what we do, combat search and rescue, and that might include getting shot at," explains Major Mike Noyes, a Pave Hawk helicopter pilot who was supervisor of flying at the 106th the night of the crash. "It takes an amazing individual to do that."

Meyer's experience was one reason his eyewitness account was so highly regarded. If any weekend warrior knew what a missile looked like, it was Meyer. It explains why, as his story evolved during the next ten days, it became the cornerstone for many theories that a missile took down TWA Flight 800.

Richardson was also familiar with missiles. "I've seen plenty of missiles. They used to shoot past us on their way to the big airplanes. They leave a big, smoky trail." On the night of the crash, Richardson didn't see anything until copilot Chris Baur called his attention to a burst of light. By the time Richardson maneuvered to a view out the front window, a tremendous explosion had already lit up the sky. He did not see what preceded it, and he did not see the smoky contrail remains of a missile.

Curiously, Meyer did not immediately share what he saw with the other two men in the helicopter. The first time anyone heard Meyer's story was when he spoke to reporters the following afternoon.

"Before the news conference I never told anyone what I saw. Because I knew what I saw. A short wait permitted me to put in order the things I'd seen and more carefully describe them," he explained.

At 4 on July 18, the guardsmen who flew the night before assembled at the front of a small auditorium used for briefings in the Air Guard's operations building. Meyer, looking fit and impressive in his blue air force uniform with a chest full of ribbons, was one of the first to speak.

"I saw something that looked to me like a shooting star. Now you don't see a shooting star when the sun is up. It was still bright. I saw what appeared to be the sort of course and trajectory that you see when a shooting star enters the atmosphere. Almost immediately thereafter, I saw, in rapid succession, a small explosion and then a large explosion."

As the days passed, Meyer elaborated even more on what he'd seen in the sky that night, his story enhanced, he admitted, by a recurring dream. In the dream, he watched airplane seats falling through the sky, overtaking other falling objects moving much more slowly. The dream was an epiphany for Meyer. He thought it signified something important about the cause of the crash and was eager to tell authorities about it. Eight days later, he joined three others who had been with him on the rescue mission at the home of Jim Finkle, the unit's public information officer.

Meyer, Chris Baur, and parajumpers, Craig "Jake" Johnson and Sean Brady were there. The men could not agree on what happened the night of the crash. Standing outside on Finkle's deck, drinking beer on an unseasonably cool and damp summer evening, they were unified in their belief that the FBI wasn't taking their accounts seriously.

Chris Baur left early. With the consent of the men remaining, Finkle called the FBI mobile command trailer at East Moriches, just a mile or two from his home, and asked for agents to come to listen again to what the men had to say.

Some time between the call and the arrival of two field investigators an hour later, Meyer and the others got into an argument. Meyer began to lecture the others about his own distrust of government and what he believed were serious inaccuracies in the history of the Holocaust. Brady and Johnson were concerned about the nature of Meyer's comments and left. Brady later told a friend that he didn't want the FBI thinking he was crazy.

Meyer stayed on to meet with the agents, but prefaced his remarks by saying, I haven't trusted the FBI since Ruby Ridge, referring to a fatal FBI seige in Idaho in 1992.

Questioned later about why he began the interview this way, Meyer explained, "The FBI is an instrument of oppression. We don't know who they are and what they're going to do with this information."

The interview proceeded cordially nonetheless. When the agents got up to leave after half an hour, there were handshakes all around. Meyer remembers that the agents thanked him for his help.

Meyer was one of more than a hundred people who reported a light in the sky just before the plane exploded. The accounts baffled investigators, until they realized they were all jumping to a premature conclusion. It wasn't what the witnesses saw that was the problem, it was everyone's interpretation of those accounts.

Psychologist Elizabeth Loftus, a professor at the University of Washington in Seattle, suggests that the witness recollections were contaminated by the rampant speculation immediately following the crash. It's only natural, after all, for people who saw something as dramatic as what happened that night to speak about it right away, to try to figure out an explanation, and to turn on the news to see what others were reporting.

Diana Weir, chief of staff to Congressman Michael Forbes, knows that firsthand. She had just arrived home from work when she was called with news of the crash.

"I jumped in the car and headed for the Coast Guard station and I put on my local radio station WLNG," recalled Weir. "Here are all these people calling in from cellular phones and they're talking about what they saw. Well over three quarters of them said, 'We thought it was fireworks, a big flash in the sky.' Ninety percent of the people who called in reported the same thing." Though these early accounts did not mention a missile, Weir quickly concluded, "My God, it was shot down."

The problem, according to Dr. Loftus, is that memory can be altered by subsequent information. "When something significant happens and there is postevent suggestion, it wreaks havoc with the memories of what people actually saw."

Still, investigators didn't need witnesses to tell them if a missile shot down the plane. If that had happened there would have been plenty of evidence on the wreckage and very likely debris from the weapon itself in the water along with the plane. The witnesses were far more valuable for what they could share about the final minute of Flight 800, but it took months for the investigators to realize that.

In the meantime, news reports being broadcast around the world were drawing the same conclusions from the eyewitness accounts as Weir had: Hundreds, even thousands of people had seen a missile hit the plane. In reality very few people actually reported seeing the plane before it exploded.

What thousands of people did see was part of the flaming descent of the 747. The majority of these witnesses were considered "latecomers," meaning that the sound of an explosion drew their attention to the sky. Since Flight 800 was flying approximately ten miles offshore and two and one-half miles high, the noise coming from the plane would have taken forty to fifty seconds to reach them. By their own admission, the witnesses had not looked up until they heard the blast. Whatever they saw had already been under way for nearly a minute.

About 180 people saw something in the sky before the plane exploded in a fireball. Of this group, only four said they actually saw the 747. On an evening as clear as July 17, the jetliner would have been small but visible. "What they would have seen is a spot, maybe with some vague shape to it," says Dr. Donald Mershon, a professor of psychology at North Carolina State University. His expertise in human perception brought him to the attention of the safety board, which was trying to figure out what the eyewitnesses saw that night.

"With significant experience identifying aircraft, someone could distinguish a large aircraft from a small one," he explained. Someone like Captain David McClaine, the pilot of the Eastwinds 737, who described the plane as a wide-body from a distance of over ten miles. He was aided by the fact that the plane had its landing lights on and he'd heard radio transmissions between air traffic control and the crew of a 747. Mershon believes most folks would just figure it was "some kind of airliner up there."

Missile experts concluded that the largest portable missile powerful enough to have downed a 747 would be ten feet long: minuscule in comparison to a 231-foot-long jumbo jet with a 211-foot wingspan and impossible to see from shore.

In conversations with reporters, Major Meyer says he believes he saw the flame trail of a missile in flight. Many others drew a similar conclusion. Yet members of the NTSB witness group were told that a missile will flame on ignition and for about the first four seconds of flight, after which the trail extinguishes and then for probably another five seconds the projectile is traveling on its own momentum, neither emitting light nor illuminated in any way.

Although the eyewitness information did not help determine the cause of the crash, it was extremely useful toward figuring out that the plane ascended after the initial explosion before plummeting back to earth in an a spray of jet fuel from the disintegrating wing tanks. The fuel mist soon ignited, causing a second, even more dramatic explosion.

After reviewing all their statements, Dr. Mershon concluded that the streak of light seen by so many people on the south shore of Long Island was not a missile.

"The vertical path probably gave the appearance of something going up toward the plane. It could have given the impression of something reaching another object and exploding. I think it was a double explosion of the airliner itself."

Both Drs. Loftus and Mershon were to have testified at the NTSB public hearing on the crash in December 1997. At the last minute, Kallstrom asked the chairman of the safety board not to discuss or disclose the witness statements. Even though the FBI had pulled out of the investigation one month earlier, Hall agreed.

Three years after the crash, the statements given to FBI agents were still cloaked in secrecy. The witnesses and the curious treatment given to their statements wasn't the only thing fueling the conspiracy theory.

Though he had been up and running nearly all night, Ron Schleede was in to work at the safety board offices early, though not necessarily alert, the morning after the crash. When his phone rang at 9, the words of the caller pierced through the foggy curtain of his fatigue.

"I could tell from his voice that it was important." Schleede recalls. Hurrying up the wide circular staircase to Chairman Hall's office, Schleede stood in the doorway and announced, I've just heard from Bud Donner at the FAA. They've got something. They won't talk to me about it on the phone. Something off the radar. I'm going over to take a look now.

It's only a short walk from the safety board offices to the FAA. Proximity belies the animosity between the two agencies. While the FAA accident investigation group is obligated to inform the NTSB promptly of anything it learns following a crash, neither Schleede nor Hall was surprised that they were hours behind everyone else looking at the information from New York air traffic control.

"They produced the famous plot, the radar dots that are going in a certain direction making a fishhook-shaped course."

Schleede remembers his first reaction.

"I said, 'Holy Christ, it looks bad.' It showed this track that suggested something fast made a turn and took the airplane."

The printout had been produced at the New York Air Route Traffic Control Center. At first, it appeared there were "radar tracks which could not be accounted for by FAA staff," according to David Thomas, the FAA's director of accident investigations. Wherever Thomas got that interpretation, it was not from his own agency's air traffic control office, according to the deputy director, Ron Morgan.

Six people from his office, all experienced at reading radar, examined the air traffic data on paper and on videotape. "We looked at it pretty closely, and we all agreed pretty clearly that there wasn't anything that was unusual," only what he described as "twinkles that pop up and go away." These twinkles, called anomalous propagation, are similar to how a television with poor reception can sometimes show snow. Morgan considered them insignificant.

Still, FAA officials in Washington determined the only prudent action was to report what "appeared to them to be a suspicious event" to the White House and other government officials.

"At this point in time, there was enormous interest on the part of the White House," explained the DOT's director of intelligence and security, Coast Guard Admiral Paul Busick. "We were not withholding information from law enforcement or the White House on anything that had significance." Sending the radar information to the White House certainly added to its perceived significance, though its meaning was far from clear.

The safety board chairman worried that the FAA damaged the future credibility of the agency by having sent information to the White House before consulting the NTSB's experts. "Obviously, that has fueled a lot of people to this day thinking there was some type of bomb and missile," Hall concluded.

"The NTSB is not the investigative arm for terrorism," countered David Hinson, the then-FAA administrator. "There is no luxury of waiting, you have to move in real time when there's potential terrorism." He added that the safety board "had the data pertinent for an accident investigation in a more than reasonable time."

Schleede confirms the atmosphere of crisis at the FAA the morning following the crash. "The FAA was working with people at the top-secret level. They were in a crisis room with intelligence people and everybody else."

Senior FAA people, along with aides to Transportation Secretary Frederico Pena, Admiral Paul Busick, Cathel "Irish" Flynn, and Kitty Higgins from the White House, were communicating on a secure videoconferencing line with representatives from the White House and the National Security Council.

"There was so much unknown at the time," Higgins said. "People were trying to get the facts. How long was the plane on the ground? Where was it headed? Were there any threats? What were the facts surrounding the flight? Sorting out this basic information was not insignificant."

In the post-crash frenzy, Admiral Busick's job was to see what U.S. military activities were going on in the area. "We thought there were DOD [Department of Defense] aircraft in the region. We went to the National Military Command Center and asked what assets were there. 'Was there anything with missile shooting capability?' I was told there was not."

Busick was also worried that FAA security might somehow have failed to protect 230 innocent travelers. "The great fear from Irish and me was that it was a terrorist attack, and for us it would have been professional failure."

Almost everybody with an interest in what had happened was going over the news as it came in, but not a single NTSB representative was even called to participate over the telephone.

The NTSB was so far out of the loop, in fact, that while its most highly paid employees were in a huddle trying to find a way to reach the crash scene, Coast Guard Admiral Robert Kramek, under the direction of the Department of Transportation, was preparing to leave for Long Island on his own fifteen-seat Gulfstream jet. Though his aircraft and the crewless FAA plane designated for the NTSB's use were sitting in the same hangar, Kramek's plane with plenty of empty seats left for Long Island hours before a crew arrived to carry the safety board's go-team.

That Thursday morning, Schleede spent less than an hour at the FAA, then hurried back to his office to let the chairman know that a missile seemed like a possibility.

"The biggest thing that scared me at the time, I told Hall, 'This isn't the last one. If this was a missile, TWA Flight 800 won't be the last one to go down.'"

Throughout the day, the FAA conducted more sophisticated analysis of the initial radar data and new information arriving from eight other centers that had recorded flight activity in the New York metropolitan area. Safety board experts were doing the same thing.

"In retrospect," Schleede concluded about the FAA's alarm to the White House, "they obviously made a mistake. I took the plot, they gave me a copy. I took it to our people who know what it means. I gave it to our people and asked them to figure out what it was."

Eventually the NTSB tossed off the anomalies in the radar as insignificant, just as the FAA's Ron Morgan had. It was a radar ghost, a ghost that came back to haunt the investigation again and again.

16

Ten weeks after the crash, in early November, Pierre Salinger was awakened at 2 in the morning by a phone call from CNN producer Ron Dunsky. Perhaps Salinger's years as a correspondent for ABC News and his service as press secretary to President John F. Kennedy had inured him forever to middle-of-the-night phone calls. Whatever the reason, Dunsky remembers Salinger was groggy but gracious when reached at his hotel in Cannes, France. It was only 8 in the evening in New York where Dunsky worked. The international cable station respects little beyond its own insatiable appetite for news.

Just hours earlier, Salinger had mentioned TWA Flight 800 in a speech to aviation executives and said he had proof that the U.S. Navy shot down the 747. The claim flew across the Atlantic faster than the Concorde.

"These are incredibly serious charges," Dunsky said to Salinger after reviewing the wire-service report on the speech. "What is the basis for your allegations?"

"I have a document here, from a high-ranking European official who got it from French intelligence," Salinger began. "I've been told it is credible. This is strong information that makes it absolutely clear that TWA was shot down by an American missile."

Salinger began reading from the document. As he did, Dunsky quickly reached for a folder from the stack of TWA notes on his desk. The thirty-four-year old Canadian was a well-respected producer in CNN's New York bureau. He'd been assigned to the crash since the beginning, so when tips like the Salinger speech crossed the wires, it was Dunsky who tracked down the details.

Something about what Salinger was reading seemed familiar.

"I'm going through the folder and I pulled out the Internet document written by the retired pilot," Dunsky recounts, referring to Richard Russell, "and Salinger's reading and I'm following along word for word."

"Mr. Salinger," Dunsky said when Salinger was finished reading. "I have this document. I got it off the Internet weeks ago."

"You're kidding," Salinger replied.

"No sir," Dunsky insisted. "Not only that, but it's been thoroughly investigated and discredited."

Salinger paused. Dunsky remembers he seemed taken aback by the news.

"I don't know what to tell you," Salinger finally said.

"Does that make you question the veracity of what you've got?" Dunsky pressed him.

"Well, that is going to make me think about it again. That may change things."

Salinger's skepticism was fleeting. By the following morning he was back in the thick of it, explaining he had more than just the Internet message to go on. The American media jumped on the story. Even CNN abandoned the judicious approach it had taken in its first reports and gave fifteen minutes of airtime to Salinger the following afternoon.

This was not the first conspiracy Salinger was bringing to public attention, as he reminded Dunsky the night they spoke. In 1990, while still at ABC News, Salinger did a series of reports on the bombing of Pan Am 103, which alleged that the U.S. Drug Enforcement Agency was indirectly responsible for allowing a suitcase bomb aboard the plane, a charge that was investigated by at least three government agencies and Congress and determined to be a hoax. Salinger suggested to Dunsky that, like Pan Am 103, the United States was hiding the truth about the crash of Flight 800.

That Salinger rejuvenated a theory that otherwise might have suffocated from a lack of substantiation, on the basis of one retired pilot's musings, infuriated the leaders of the investigation. The whole mess was creating a lot of anxiety for people who lost loved ones in the crash and for the sailors who'd spent nearly five months retrieving victims and wreckage. It also discounted the work of metallurgists and engineers and scientists and pathologists who'd gone over the wreckage and the human remains without finding the merest hint that a missile had been involved.

Given his stature as a former high-ranking government official and a journalist, and the international press attention his comments were getting, Federal authorities realized they would have to respond to Salinger's claims.

That afternoon, in the large and crowded briefing room at FBI headquarters in lower Manhattan, Jim Kallstrom; Jim Hall; Brian Gimlett, special agent in charge of the Secret Service in New York; and Navy Admiral Edward Kristensen, who was in charge of the navy's effort to retrieve Flight 800 wreckage spent more than an hour explaining why Salinger was wrong. And not just wrong. Kallstrom said the charges were "absolute, pure, utter nonsense."

"I can assure you we've looked at every angle, every possibility, and the military of this country has had nothing to do with this horrendous tragedy," Kallstrom continued. Five days later, challenged by Kallstrom that if he had something to share with authorities he should do it, Salinger arrived in New York. Kallstrom's deputy Tom Pickard was sent to met with him at a midtown hotel the evening of November 12.

What do you have that we can move on in this investigation? questioned Pickard as the men sat on the sofa in Salinger's suite at the Parker Meridian Hotel.

"Its all there on the Internet," Pickard remembers was Salinger's vague reply. "You guys are the investigators. Look at it."

What details do you have? What is the overwhelming evidence? Pickard asked, frustrated at Salinger's lack of answers. After an hour Pickard left. "The whole thing was bizarre," he said. "He wasted my time."

Salinger had become a magnet for assorted conspiracy theorists. The debate, simmering in the anonymity of the Internet, was now at full boil with a celebrity spokesman. Salinger gave credibility to the Internet chatter and the cyberspace residents, like Ian Goddard, who sustained it.

Right after the crash, Goddard, a thirty-five-year-old graphic artist, assiduously reviewed every bit of media coverage and created an impressive Web site that was a patchwork of news reports, smart-looking graphics, eyewitness quotes, and personal commentary.

Goddard's work impressed Salinger. The artist was delighted to become part of his investigative team that included an old friend of Salinger's, Mike Sommers. According to Goddard, "Sommers said he and Pierre went way back to the Kennedy Administration."

The three prepared a sixty-nine-page document, much of it from Goddard's Web site, the rest a litany of charges by unnamed sources and reports of sailors who witnessed the event but would not agree to be identified.

By the spring, Salinger and his team had hooked up with Richard Russell, who, like Goddard, was happy to have someone of Salinger's stature helping to "get the information out, one way or the other." Russell's contribution to the group was a videotape of the air traffic control data that had so intrigued the FAA the night of the crash.

Stills from the videotape were published in a popular French news magazine, *Paris Match*, along with a story bearing the byline of Goddard and Sommers.

After that, Russell traveled with Sommers to London, where the two met with staffers at ABC's London bureau. Russell says he was horrified to hear Sommers offer the videotape to ABC for $1 million. Afterward, he says he thought he needed to warn Salinger that Sommers's motivations seemed different from their own. "It's a big mistake for you to be with him," Russell says he told Salinger. "He's not doing you any favors and he's doing you a lot of harm."

An ABC spokeswoman says the network turned Sommers down. Russell was no longer associating with Sommers when a similar, more subtle, often rambling proposal was put to news executives and a producer with the Larry King show at CNN.

"Larry could show the tape and you guys can . . . I have the rest of the story which is horrifying and I don't think Larry will want all the aspects of that story," he said, and, referring to the network's news division, "and you guys can have the rest of it. Yeah, provided Ted pays some money to TWA 800 victims."

What kind of money was he talking about?

"I don't know. You have to make an offer. Ted will know," Sommers replied, speaking of CNN founder Ted Turner. "It will make a hellava movie and a hellava book and Ted will make millions on it."

CNN passed on the Salinger exclusive.

17

The south-facing windows in the NTSB offices provide an expansive view of planes taking off from nearby Washington National (Reagan) Airport. Bernard Loeb, the prickly director of the Office of Aviation Safety, had positioned his desk with his back to the view to avoid distractions. He had all he could handle in the summer of 1996.

Loeb's department was burdened with several high-profile and troublesome crashes, two of which appeared to be deadly repetitions of earlier accidents. The cause of the crash of USAir Flight 427 outside Pittsburgh in 1994 was frustratingly elusive. It looked likely, though, that the popular jetliner had a design flaw in its rudder system.

The safety board had already determined that ice on the wings brought down an ATR-72, a French-built turbo prop, over Roselawn, Indiana, in October 1994. Sixty-eight people had died. Though the board had issued a final report, indicting the FAA and its French equivalent, the French Directorate General for Civil Aviation, for failing to act on known icing problems on the Aerospatiale airplane, staffers knew they had not heard the last of American Eagle Flight 4184. There was talk that the FAA was going to take the unprecedented action of appealing the board's finding of probable cause. It was one of the more controversial of the board's investigations, engaging it in interagency disputes and international politics.

Still sitting in piles on dozens of desks at the safety board were the reports from the Valujet DC-9 crash of May 1996. Once again, it appeared that larger issues were involved, including questionable FAA oversight of discount airlines.

These were some of the responsibilities on Loeb's mind on the evening of July 17, 1996, and he uses them to explain why, when he was called with word of the Flight 800 disaster, he did the one thing experienced crash investigators never do. He made assumptions. From his kitchen counter, with the television tuned to news of the crash, Loeb jumped to the conclusion that a 747 exploding in-flight was very likely brought down by a bomb. In that case, he reasoned, solving the case would be the FBI's problem. As it turned out, the crash was his problem, and so was the FBI.

The Grumman Aircraft manufacturing plant produced more than a thousand fighter jets for the U.S. military between 1957 and 1996, from a complex of hangars and runways on nearly 3,000 acres in Riverhead, Long Island. It was seventy-five miles and a world away from New York City.

When Grumman merged with Northrop to create the Northrop Grumman Corporation, control of the property was returned to its owner, the U.S. Navy, where it provided employment to about twenty-five people who worked to secure and maintain it. For the most part, it collected dust and animal droppings.

It was the perfect place to take the remains of TWA's jumbo jet. The local road that provides the only access to the plant is not heavily traveled, yet it is capable of handling the eighteen-wheel flatbed trucks that would be arriving around the clock, loaded with the remains of the plane.

Friday following the crash, the vacant plant snapped back to life and arriving investigators soon outnumbered the deer, beaver, and geese who had taken over in the fields outside the buildings and the cats who'd laid claim to the hangar. Interspecies relations were fine. Among the humans it was another story.

Beyond the security gate is the first and largest of a dozen hangars on the property. At more than 300,000 square feet it was big enough to accommodate the reassembly of a ninety-foot section of the 747's fuselage and some of its internal structure. This is where most of the investigators would spend their waking hours, in the white noise of humming fluorescent lights, whirring ventilation fans, and metal shards being moved around. The smell was a blend of new carpet, old fish, and charred wreckage.

A second, smaller hangar was for the reconstruction of the aircraft cabin, complete with galleys, bathrooms, and passenger seats. Two towering heaps of unidentifiable cabin material sagged by the wall, making this hangar a poignant venue for even hardened investigators. The people who worked here were not hardened at all. In fact, several in the cabin interiors group had never been on a crash investigation before.

The overwhelming task of examining the wreckage and putting it in order, combined with the horror of the event, partly explains why tensions among the various agencies ran so high. Differences in culture between the necessarily secretive crime solvers and the habitually open safety board team were also difficult to bridge. But the basic power disparity between law enforcement and everyone else probably played the largest role.

The FBI agents arrived at what they considered the crime of the century. They were full of bravado, boasting of their success in bringing to justice the suspects in the World Trade Center bombing.

The NTSB investigators were no less confident, but here they were on unfamiliar ground. They'd all worked cases in which the FBI had an interest, including the investigator in charge, Al Dickinson, whose last major crash had been United Flight 585, a 737 that went down in Colorado Springs in March 1991.

"The FBI was there. They came out with two agents," Dickinson said, marveling at the difference between that investigation and this one. No NTSB investigator had ever seen a law enforcement turnout like this, which explains why they yielded to the FBI early on. They were also following the instructions of Bob Francis.

On the flight up to Long Island, Francis made it clear that the investigators should cooperate with the FBI. "One of the things that Francis mentioned was that we should not cause any issues with them," Dickinson said. "We should try to work out everything with them."

One consequence of this early go-along and get-along attitude is that the FBI conducted witness interviews without including NTSB investigators. The FBI even excluded the NTSB investigators from talks with the ground and maintenance workers at JFK, whose comments would be highly relevant to the safety board's investigation.

Francis insists he never gave Kallstrom the go-ahead to do this, but Kallstrom remembers it differently.

"I said to Bob the first day, they were invited to go on any interview they wanted, but they had no one to go."

Francis was also taking heat for not being fully aware of what the tin-kickers were doing. With some regularity, news reporters had information before he did.

Francis acknowledges that moving by helicopter from Navy ship to Coast Guard station to the hangar at Calverton to JFK assured he would often be hours out of date. In addition, the news conferences, scheduled to accommodate reporters' deadlines, meant that he would also miss the nightly progress meetings where the latest developments are announced.

Unable to keep up with the far-flung investigation, Francis had taken to spending his days with the FBI boss, counting on

Kallstrom's much larger staff and much better communications to stay up-to-date.

When Hall, the NTSB chairman, felt the lack of communication had gotten out of hand, he offered to come to Long Island and help Francis out.

Francis said no.

"Well, what I did then was, I just eventually went up later," Hall said. "I went up to the hangar and saw Kallstrom."

Now the chairman wasn't coming to help; he was coming to reestablish turf.

On Sunday afternoon, July 21, her second day with the families, Tonice Sgrignoli broke a promise she made to herself and was drawn into a conversation with a woman her same age in the family group. She had scrupulously avoided any personal contact with the families. The self-imposed isolation allowed her to justify her impersonation because she was not overtly lying or putting herself in a position where someone might confide in her. She was there as an observer, she told herself, though she was inexplicably sad, as though the grief of others had infected her, too.

Johanna O'Flaherty never met Sgrignoli, but finds the description of her state of mind perfectly understandable. "There's a thing called vicarious traumatization. It's actually becoming the container for the grief of the people who are grieving themselves," O'Flaherty explained. "It's intoxicating and incredibly powerful."

Sgrignoli and the woman wound up spending the rest of the afternoon together. The following day they rode to another memorial service in Long Island.

The families were escorted by police cars and motorcycles. At the end of the route, hundreds of residents lined the road, holding signs reading "God Bless You" and "You Are in Our Prayers," a stirring display of sympathy that remains a cherished memory for many of those who witnessed it.

Cynthia Martens, whose daughter lost her best friend, Larissa Uzupis, in the crash, wrote that the "compassion of those people, more than anything else, gave me hope and encouragement."

Sgrignoli was also moved. "I could have lost my identity in this process," she said.

The mourners were seated in chairs that had been set up facing the ocean on a wide expanse of white sandy beach. The service was stirring, with music and prayers in three languages. Toward the end it turned heartbreaking as mourners spontaneously removed their shoes and waded into the ocean fully dressed, gripping long-stemmed roses and each other.

Sgrignoli wanted to file a story with her newspaper that described just how powerful the event had been, how it had driven the grief-stricken to impulsivity. She wanted to tell readers how "the sea smelled like roses."

"I cared a lot about these families and wanted to protect them," she realized. "Even though part of what I wanted to protect them from was myself."

After returning to the hotel that afternoon, Sgrignoli called her editors at the *Post* and gave them the details from the day. She felt anxious that she could not distinguish herself as a reporter here because she could not conduct interviews with officials without revealing herself. She was also discouraged because the subtle details of the emotional drama going on around her were not making their way on to the pages of the rock-'em, sock-'em tabloid.

"That was my growing frustration of being in this double role; immersed with the grieving families and feeling obviously like a failure, and a failed reporter because I didn't think my coverage made a difference," Sgrignoli explained, "because in no way, shape, or form could I justify what I was doing in terms of the result."

Feeling that nothing significant had been accomplished by her presence in the hotel, Sgrignoli decided to leave that evening. She called over the escort assigned to the woman she'd come to know and revealed her situation to him. Sgrignoli thought he would explain her situation to her friend, so that her disappearance would not be a mystery. She was sure he would understand.

He did not. Ten minutes later the reporter was being led from the hotel in handcuffs.

Jim Kallstrom was not surprised at what he was hearing from Jim Hall of the NTSB. During the intense days of the investigation, Kallstrom and Bob Francis had gotten to know and trust each other well enough for Francis to share that his own agency was not happy with his performance. "I realized he was getting beat up mercilessly by his own people," Kallstrom said. Hall's visit to Long Island reinforced the point.

"Jim, I just want to let you know, on this investigation Al Dickinson is the investigator in charge," Hall said to Kallstrom, handing him a photocopy of the agency's organizational chart.

"Gee, Jim, I think it's kind of strange you're telling me this at this point in time."

"Well, we should have told you earlier," Hall said, pausing for Kallstrom to respond.

Kallstrom was being disingenuous. Ken Maxwell, the senior FBI agent at the hangar, had already passed on a similar message he'd received from the NTSB through investigator Ron Schleede. Yet the domineering Kallstrom considered the relationship he forged with Francis to be working just fine. He would put off forever dealing with anyone else from the safety board if he could. If Francis was claiming authority inappropriately, the NTSB should work it out.

Kallstrom was thinking these things privately. To Hall he simply said, "What do you want me to do about it?"

"I just want you to understand how it works," Hall replied.

. . .

For the first six weeks following the crash, the suggestion that the FBI would soon take over the investigation was made many times. Kallstrom announced just two and a half hours after the disaster that the FBI would be investigating it as a potential criminal destruction of aircraft. Two days later, unnamed FBI sources were telling CNN the agency was "poised to take over the investigation." Two weeks later, special agent Joseph Cantamessa told the *New York Times* that "one positive result" from lab tests on chemical residues would be enough, while the *New York Post* headline read: "Feds Are All But Certain of a Bomb."

It was understandable that law enforcement might jump to conclusions. So much sounded suspicious. If it was a crime, it was a monumentally audacious one. Cops would have to sprint to catch up with the perpetrators.

The public, the press, and the politicians could be expected to think the worst also. The Olympics were under way. A terrorist was on trial in New York, accused of trying to blow up airliners. What was less explicable is why people who should have known better surrendered to the speculation.

Bernie Loeb was one of them. Touchy and arrogant, the balding, perpetually weary-looking chief of aviation investigators had been in that job for just eleven months. People often describe him as one of the smartest men they know. But for all his brains, he had the same pedestrian thoughts as the neophytes the night of the crash. Listening to his investigators on the scene forced him to rethink his position.

"Crimes are corroborated fairly quickly. Organizations take credit, law enforcement hears something, the wreckage reveals something, the CVR gives clues. We weren't getting anything to suggest that a criminal enterprise had taken place," Loeb explained. "No evidence from the floating debris, no pitting or gas washing on the metal. The bodies gave us no evidence, no fragments or materials. There was no shred of evidence telling us this could be a crime."

The significant difference between how law enforcement and air crash detectives approached the investigation is illustrated by the interpretation of this lack of evidence. Though it did not affect the firmly held opinion of crime fighters that the disaster was caused by a bomb or a missile, it sent the NTSB looking elsewhere for a cause. "Once we started getting that wreckage in, we started to think, 'this looks like an aircraft accident,'" said Schleede, a thirty-year veteran with the NTSB, known for his firmly held opinions. "Despite the impressions left by the media, we weren't surprised to see the things we were seeing."

Radar data and eyewitnesses established that the plane had broken apart in flight. Acting on the maxim that there are no new crashes, the question became: When have we seen this kind of rapid in-flight breakup before?

They'd seen it in 1988, when eighteen feet of the top of an Aloha Airlines 737 peeled away, the result of structural failure, while pilots miraculously maintained control of the aircraft enough to land it safely.

In-flight breakups had also been caused by fuel tank explosions. The wreckage from the TWA 747 supported this scenario.

"We started getting pieces with fire damage, obviously a fire here . . ." Merritt Birky said, pausing to recapture his reaction at the time to the first center fuel tank pieces that he inspected. "Right from the start, a fuel-air explosion was considered a possibility."

Birky had seen that before, too, in the explosion six years earlier of the center wing tank of a Philippine Air Lines 737 in Manila. Eight people were killed. The people seated over the fuel tank had their ankles broken by the force of the floor lifting up with the pressure of the blast.

Standing in the hangar in early August, he was explaining to systems group chairman Bob Swaim how alike the events were. The tanks were similarly pushed out at the top.

"I remember pointing out to Bob the similarities between the top of the tanks. The tops were bowed in a similar fashion in the front of the tank. That told me there was a fuel-air explosion. The top was bowed because of the over pressure the explosion created."

There were differences, too.

On the 737, the center tank capacity is roughly 1,600 gallons. The 747 center wing tank carries eight times as much fuel in a far more complicated structure consisting of nine separate chambers, six access panels, and a floor-to-ceiling dimension that declines from six feet at the front to four feet at the back.

I'm not surprised that there could be a fuel-air explosion with the volume of energy in a tank this size, Birky told Swaim.

What kind of temperatures can we get in here? Do we get above the lower flammability level? Birky asked, more to himself than anybody else. Let's get the numbers from Boeing.

At forty, Bob Swaim resembled the 1960s TV character Beaver Cleaver. His personality is youthful, too, friendly and engaging on a good day, cross and short-tempered when things are not going his way. An eager tenaciousness makes him unlikely to anticipate or tolerate failure.

When Birky began to list the kinds of information he thought the tin-kickers would need to zero in on the fuel tank, Swaim enthusiastically added something of his own, arranging a field trip to the inside of an intact 747 center fuel tank.

Twenty years separated Monica Omiccioli from her mother's brother, Jean-Claude Poderini. Monica's choice of lifestyle, traditional and religious, was a contrast to that of her ponytailed uncle. Jean-Claude had never married and probably never would. Yet the two were more like friends than relatives. He was her confidant and mentor at the design house where they worked. She was his link to contemporary thought and taste.

She was also the person with whom he talked about death, and always at her instigation.

"Monica was resigned to die young and her lifestyle reflected her feelings about dying young," Poderini said. A popular love song on the radio referred to a couple together for twenty years and she would say, "I'm never going to have a twenty-year anniversary because I'm not going to live that long."

"Her lifestyle was active and vivacious, but she would get melancholy with no prompting," Poderini said.

So when days passed and neither her remains nor those of Mirco Buttaroni had been identified, it was Poderini who stayed on at the Ramada, while the rest of the family returned to Italy, to wait together in the weak comfort of familiarity. It was Poderini who delivered dental records and strands of hair from Monica's hairbrush to the medical examiner's office. In the beginning of August, it was Poderini who delivered the couple's wedding clothes to a Long Island funeral home so the two, found at last, could be dressed for burial.

The bodies of Monica and Mirco had been retrieved from the ocean on July 31 and were returned to Italy together on August 9.

That same morning, a dozen investigators arrived at a TWA hangar at JFK where a 747 in for routine maintenance was parked. Beneath the massive undercarriage, a steep ladder was topped with a work platform on which stood a small air compressor with several rubber hoses and ventilation masks attached. This compressor would provide fresh air for those going inside the tank.

The Boeing 747 had dominated the thoughts and conversations of these men for weeks. The one they'd seen was tattered and pathetic, completely unlike the remarkable and lovely airplane before them now. Michael Huhn, an engineer for the Air Line Pilots Association, found the contrast startling when he made a similar trip to TWA's hangar the following month.

"I remember standing there and looking at this huge machine and thinking how little we had in the hangar at

Calverton. The drama was the difference between the flying machine and the mess."

Within the confined space of the hanger, the jet's size was even more impressive. Huhn was excited about entering the center tank, considering it a way to become "more intimate with the airplane."

The investigators were instructed to empty their pockets to avoid the unlikely possibility of creating an electrical charge that could ignite the vapors in the tank. They tossed hotel keys, pagers, cell phones, pens, rings, and watches onto a table by the wall, patting themselves down for anything they'd overlooked.

When it was Swaim's turn, he climbed up the ladder, knelt on the platform, and pushed his head and arms through the eighteen-inch opening at the far corner of the tank's front exterior wall. He drew his torso through by grabbing an inside floor beam with his hands and pulling himself forward. When his legs were inside the tank, he stood up. Moving in here was going to be tricky. With each step, he risked being tripped up by a grid of structural supports along the floor.

Swaim entered through an area called the dry bay, which does not carry fuel. He didn't stay there long. "It was not where the action is," explained the NTSB's Birky.

This was not the first fuel tank Swaim and Birky had investigated, just the biggest. Swaim realized that the moment he extended his full six-foot, one-inch frame. That he could stand at all was remarkable. When he'd needed an inside look at an airplane fuel tank on other crash probes, he was accustomed to snaking around on his belly.

The walls and ceiling in the 747 tank were shiny, new-looking, and slick with a film of jet fuel. Small puddles on the floor made each step even more hazardous. The smell of kerosene seeped through the face mask.

Unlike the others, Swaim had not given up his notepad or pencil. He struggled to hang on to them as he tucked himself through the small access doors dividing the tank into increasingly smaller compartments. His flashlight cut a weak beam through the vapors.

He began a catalog of what he was seeing: the fuel probes braced against the walls, how the wires connected to them, and how the wires themselves were pinned to the wall.

At the back of the tank, the ceiling was too low to allow him to stand. Here, it was oppressively narrow, hot, dark, and airless. He sank onto one of the floor beams and silently studied the structure. Slight metal bonding strips attached to pipes where they coupled. Were they all in place? Where were the fuel pumps mounted? What was their proximity to each other and the fuel lines that ran along the back of the tank? He could get the specific numbers at the hangar from the Boeing representatives, but he felt a need to read the tank, assess its shape and structure firsthand if he was going to get to the bottom of the mystery of Flight 800.

18

By the time Los Angeles–based FBI agent Rich Hahn arrived at Calverton in mid-August, interagency resentment was full blown. The tin-kickers, having come to the realization that this crash might very well be an aviation issue, lost their tolerance for the FBI presumption that a crime had been committed, thinking it rash and unfounded. They were offended that the agents controlled the hangar and the wreckage, and showed no interest in their opinions.

Bob Swaim remembers seeing several FBI agents, including supervisor Ken Maxwell, huddled over a beam with a large hole in it. Without being asked, Swaim piped up that hole was where the wing landing gear had ripped out.

No, Swaim remembers one of them replying, it still could have something to do with a missile.

Go ahead, he said. Swaim sighed and turned to walk away but in his pique he couldn't resist having the last word. I don't know why you guys want to worry about things we can identify. It's normal.

Normal? Now it was Maxwell who was frustrated. He stood up and gestured across the vast space with open arms, replying, None of this is normal. Airplanes don't break up like this.

Ken, I just came here from Valujet, Swaim replied. No piece of that plane was bigger than a dinner plate.

Later, Maxwell insisted, "We relied on their expertise. We went to them all the time." But he thought the safety board's answers were sometimes incomplete. "We don't just settle, and say, 'Hey, that could be it. Yeah, it's probably that, let's go home.'"

Into this simmering discontent, Hahn was introduced with little fanfare but much deference. Fellow agents talked about how Hahn's sleuthing skills led to the determination that a bomb had been placed aboard a Boeing 727 that crashed in Colombia in 1989. A special area on the floor near the fuel tank was set apart for him to work. Others were asked to stand aside and "let Rick do his thing."

His first stop was the scaffolding on which some of the retrieved center fuel tank wreckage had been lashed. Toward the back end of the tank a wall called span wise beam two was heat damaged and wrinkled, making it look like a curtain drawn open.

This piece had intrigued some folks in the hangar since the day it arrived. To them it suggested the pathway by which a missile entered the tank. To make that case, Hahn would have to find where it first penetrated the plane. It looked like it would have pierced the right wing before hitting the center tank. If the damage was there, Hahn was going to find it.

His plan called for rebuilding part of the front of the right wing. Some investigators called it busy work, but Bob Benzon, who was filling in as the investigator in charge so Dickinson could get a few days off, didn't see the harm in it, and neither did Boeing's chief investigator, Dennis Rodrigues.

Put it together and see if there are any openings there, Rodrigues said with a shrug. When the work was done, the result didn't surprise him. There weren't any holes lining up with the wrinkled piece. Rodrigues knew the wreckage would tell the story.

Hahn was still hard at work on August 17 when David Hinson came to the hangar. The FAA administrator had flown himself up to Calverton in a hurry after Francis called and confided in him that the FBI was considering declaring the crash a crime and taking over the investigation.

Francis trusted the opinion of the FAA chief, a sixty-four-year-old former Navy pilot. That Francis would turn to the FAA for advice, over his own people at the safety board, was another irritant at the NTSB, which relishes the distinction between the two aviation safety agencies and its own scrappy independence.

In a small room on the second floor of the hangar, Hinson joined Francis and Kallstrom in a long conference call with FBI chief Freeh.

The FBI lab in Washington had two positive readings for chemical explosives on wreckage of the 747, Hinson was told.

You say you found residues of chemical explosives on the lab tests. What does the evidence on the wreckage say? he asked Kallstrom.

There isn't any yet, Kallstrom replied.

What about forensic evidence from the passengers seated nearby, anything there? he queried.

Nothing, but we're still checking.

And on it went, the four men analyzing what they knew, comparing it against what they did not. Finally, Hinson turned to Francis, who had originally asked for his advice, urging him to wait before making a decision about whether the investigation should be turned over to the FBI. You cannot cross this line until you know for sure, he said. If it has to be undone later, it could be very difficult.

At the same time, on the floor of the giant hangar, Merritt Birky went over with Rick Hahn all the reasons he did not think the wrinkled section of span wise beam two was evidence of a missile.

This deformation was made after the piece hit the water, he said, leaning into the scaffolding and outlining with his arm the folds in the aluminum.

See—he gestured again—this piece got hot from the fire prior to impact with the water. You can tell because the heat damage is uniform throughout the wrinkles. If there had been an impact with a missile, that would not have been the case. There's no scar mark where a missile would have impacted.

Looking up, he saw Kallstrom approaching. Whether he was worried or agitated, Birky couldn't tell, but something was clearly on his mind.

Merritt, Kallstrom called out in his flat Boston accent, lowering his voice and guiding Birky away from Hahn. Have you been told about the explosive residues we've found? I think it's evidence there was a bomb on the plane.

A bomb? Birky repeated incredulously. No. I haven't heard this.

Well, we've found traces of plastic explosives. RDX, PETN. Roger's done the testing in Washington, referring to Roger Martz, the chief of the chemistry unit at the FBI lab headquarters in Washington, D.C.

Kallstrom waved his hands dismissively as he spoke, as if there was nothing more to be said about the matter, but he'd sought Birky out precisely because he knew him to be cautious, and Kallstrom wanted his perspective. For once, Kallstrom noticed Birky wasn't arguing.

Behind the uncharacteristic silence, Birky's temperature was rising. He knew Martz and respected his work, but did not think the FBI lab in Washington practiced good science. Birky's misgivings would soon be ratified by allegations of mishandling of evidence and contamination of exhibits there. Three of the FBI scientists working on the crash, Martz, Rick Hahn, and James (Tom) Thurman, an FBI explosives expert, would be accused of practicing shaky science or testifying at criminal trials beyond their expertise.

Several years before, Birky made a decision not to use the FBI lab unless a duplicate analysis was being done elsewhere. Here, he'd been excluded from the testing all together.

This investigation was like no other in which he'd participated. He'd resolved two or three times already to bite his tongue before the politics of the thing made him crazy. He looked into Kallstrom's determined face and replied simply, "I don't believe it without seeing the data myself."

Fine, Kallstrom answered without hesitation. I'll arrange for Roger to meet you at the lab. Time's of the essence, though. We're sure it's a bomb, which will obviously affect the investigation, so get on this right away.

You bet I will, Birky answered, turning away from Kallstrom and heading back to the fuel tank. He was confounded. There were no physical signs consistent with a bomb exploding anywhere near the tank. Where had those residues come from? What did they mean? Birky started rearranging the upcoming week's schedule in his head to accommodate an immediate return to Washington.

On a table covered with a sheet of protective glass, in a suite of rooms on the sixth floor of the Ramada, detailed photographs of earrings, watches, rings, and wallets were laid out for the families to view. These items had been found with the bodies of the unidentified dead. The medical examiner's office had inventoried these personal effects and was showing them to the family members in hopes they would recognize some things.

At her wedding, a priest who had known Monica Omiccioli for years had presented her with a ring, a gold cross shaped into a circle.

When her body was found, two weeks to the day after the crash, the ring was still on her finger, helping officials put a name to her remains.

The medical examiner was using X rays and dental records, DNA and blood samples, methods both scientific and logical.

Laying out sentimental items this way was not the only delicate decision Dr. Wetli had to make. The team of forensic examiners and dentists were having a difficult time identifying three women, so they photographed their faces and offered to show them to those awaiting the recovery of women of that same general age.

Wetli's office was under fire from many sides. The families felt he was moving too slowly in releasing their loved ones for burial. Politicians blamed him for getting a late start and refusing offers of assistance. To many in Suffolk County, including Wetli, this was the fallout from having the families so far away.

"I wasn't able to talk to the families directly. I was doing it by the media. I couldn't be there and here at the same time. It was a nightmare," Wetli explained.

Often when a body arrived or medical records were needed, a one-and-a-half- to two-hour drive from the morgue to the Ramada was required. Detectives, forensic technicians, and funeral directors were spending half of their day driving. TWA Ambassador Club hostess Mary Anne Kelly made the drive a few times to pick up medical records pilots were bringing in from Europe or material from the TWA escorts at the Ramada.

After the frustration of being at the hotel the night of the crash, Kelly had agreed to help out at the medical examiner's office, though nothing in her training at TWA prepared her for it.

Most days, she stayed at the morgue in Hauppauge, sitting at a table that ran down the middle of a conference room, lined on either side with more tables for the forensic dentists and detectives.

A mosaic of photographs had been pinned to the wall, giving faces to the victims. The job remaining was to give them names as well.

Autopsies were performed in the adjacent operating room. When a body was identified, a detective would announce it to those in the room. It happened more than 100 times, but each time, Kelly remembers, there was a momentary halt in the activity. "We would stop and pause and reflect and look at the picture on the wall. Sometimes we would see where they were seated according to the manifest and the seating chart."

If the victim's family was staying at the Ramada, a call would be made to the detectives there so they could notify the family in person.

Wetli's assistant, Bob Golden, had moved into the Ramada. His people skills were everything Dr. Wetli's were not. Self-effacing and empathetic, Golden could quickly assess what news a relative would want to know and what was best left unsaid.

Dr. Wetli, by contrast, was obstinate and reserved, a lightning rod for the anger of the families and an easy target for publicity-seeking politicians. Families were incredulous that Wetli turned down an offer by the governor to send more personnel to the medical examiner's office. And Governor Pataki was just plain angry. Wetli's bosses were called, including Suffolk County health commissioner Mary Hibberd.

"I got a call late one evening to get to the medical examiner's office. The governor was so angry he was going to move the bodies to the Armed Forces Institute of Pathology." The Armed Forces uses a mortuary at Dover Air Force Base in Delaware for processing victims of mass disasters. On occasion it is used to help in civilian matters. Hibberd was concerned by the threat. "That would be a big mistake," she told her bosses. "The lab was going to do a good job."

Wetli explains he never turned down help that was useful, but people were being thrown at the office. "The call I got from the New York State Emergency Management Service was, 'We're gonna send some help.'" When Dr. Wetli pressed for information about who would be sent and what kind of specific skills they could offer, he claims he was told, "We don't know yet, it depends on who calls in."

It wasn't people that the medical examiner needed, anyway. He had forensic pathologists and dentists at the ready. He needed information, X rays and fingerprints and blood samples that could be used to compare against the victims. At the Ramada, efforts were being made to get that material sent to Suffolk County from around the world. Family members were asked to answer questions about the deceased.

The questionnaires triggered yet another controversy, though, because of the nature of the questions. If the victim was a smoker or had heart disease, it allowed pathologists to focus on remains that showed evidence of smoking or heart damage.

Meanwhile, Mayor Giuliani was warning the families that in the inevitable lawsuits, this information could be used to reduce the awards to the survivors. His staffers distributed the medical forms to reporters and urged them to question the intentions of TWA and its insurers. A lawyer from the Queens district attorney's office was called in to advise the families to be careful.

Colonel Dennis Shanahan, commander of the Army Aeromedical Research Lab at the time, had his own idea of what Wetli's office could have used.

"The support he needed was for his bosses and the politicians to support the concept that this may take a little time," Dr. Shanahan explained. On loan from the army, he was helping the FBI and the NTSB on the investigation. "It has always been my experience that if things are explained correctly to families, they will support an excellent investigation over a rush to release the bodies. As it occurred, the only thing the politicians and even the NTSB pushed was to get the identifications done and get the bodies back to the families."

One great asset at Wetli's office was a seventy-five-member team of forensic dentists from Long Island. These men and women attended professional seminars and worked and socialized together for years before the crash. Their familiarity with the office and one another enabled them to start making dental identifications the minute the records arrived.

This progress was hardly even noticed during the fractious meetings of families at the Ramada. Dr. Wetli's boss, Suffolk County Executive Robert Gaffney, believed the families were being politicized, "spun" as he described it, by people who were unaware of what was really going on. As for the medical examiner, the process "took a toll on him but he functioned beautifully," Gaffney said. "Wetli is a hero in this process."

For four weeks, Mary Anne Kelly worked in this stressful and emotionally charged atmosphere, out of sight, out of her element. Few new faces came to her part of the building, though Kelly does remember meeting the police chief of Montoursville who was being given a tour, and Jeffrey Erickson and Johanna O'Flaherty, who visited to make sure Kelly was handling the pressure.

O'Flaherty brought an aromatic mix of eucalyptus and pine in a small bottle, something Kelly could dab under her nose to counter the odor of the morgue. Smells can be a powerful trigger of memory and O'Flaherty hoped aromatherapy would help prevent some of the more difficult ones she was sure Kelly would have in the future.

"Little things like this in the aftermath of the disaster can really help," O'Flaherty reasoned.

In the middle of August, when nearly all the bodies had been identified, Kelly left Suffolk County and went home to her family in Queens. "Working out there you look at things differently," the mother of two explained a year later. "You bless every day of your life and thank God for waking up."

Raju Shangvi, forty-five, heard the news that Hank Gray had died on Flight 800 and did not need to be reminded of the promise he'd made to Gray. The two men had discussed it often over the course of their seventeen-year friendship.

The most recent time was in the summer of 1994, when Gray, Shangvi, and Elena Barham were in Arizona. The three had been out having a few drinks and were walking back to their hotel rooms when Gray, the tallest, slung an arm over each of them and drawled, Ya'll are my best friends in the world. He sounded slightly tipsy.

Turning to Shangvi, he continued, Indian, when I die, promise you're gonna come get me and bring me home.

I will, Bubba, Shangvi said, laughing indulgently.

On July 27, Shangvi went to a Long Island funeral home and asked to see the remains of his friend. He was taken to a back corridor where the coffin had been pushed up against the wall. The mortician had tried to talk Shangvi out of viewing the body, but Gray's friend was insistent. It was part of fulfilling his commitment.

Alone by Gray's coffin he did not cry, though he felt a powerful sadness. "I didn't believe what I was seeing, that it had happened," he recalled.

"I promised you," Shangvi said, addressing Gray's corpse.

"If you wanna go, let's go."

President Clinton's trip to New York was an on again/off again thing. An advance team appeared at JFK three days after the crash, checking out TWA's Hangar 12 where the Port Authority was organizing a closed memorial service for Sunday afternoon. John Cardinal O'Connor, archbishop of New York, was going to be officiating.

When the presidential trip did not materialize, some suggested that O'Connor's differences with the president over his veto of a bill outlawing some late-term abortions was responsible. O'Connor made it clear to people planning the event that he would not attend the service if Clinton was there. The tentative plan for the president to attend the Sunday service was abandoned.

Events at the family hotel, meanwhile, were getting uglier and more public. Angry family members had lashed out at TWA, then the medical examiner. The federal government might be next.

At the White House, Kitty Higgins described the reaction. "The feeling was we had to do something to show that we cared. And we had to change the conversation." A trip to New York was put back on the president's schedule.

From the night of the crash, Higgins had been advising the heads of the agencies involved to watch what they said and how they acted. "Valujet was a public relations disaster for the FAA and wasn't handled well from the standpoint of the passengers," explained Higgins. "When TWA happened, I knew we had to be very hands-on in the response."

Still, if anyone expected the president to arrive in New York and talk turkey with the battling agency heads and interfering politicians, they were disappointed. In a thirty-minute meeting at JFK on July 25, Clinton got updates from the leaders of various organizations. The president thanked them for their efforts. Then he posed for pictures with each of them before leaving for his meeting at the Ramada, where his famous listening skills turned the tide.

Clinton was joined by First Lady Hillary Rodham Clinton in the Ramada ballroom. He told the families that the recovery of bodies would be the top priority. One week after the crash, so many were still waiting to hear that their loved ones had been retrieved. He also urged them not to speculate about the cause of the crash.

It was at this meeting that the president announced the NTSB would be the agency in charge of meeting their needs. This was good news. The families liked Bob Francis for his understated compassion. Having lived in Paris, he was prone to speak to the French in their language. They were equally fond of Peter Goelz, whom they could count on to be straightforward and sensitive. Though he was always in a hurry, he never made the people he dealt with feel rushed.

The president kept his talking to a minimum, inviting those in the room who wanted to speak to him to do so. For three hours they did.

Spouses, parents, and lovers pulled out photographs and sentimental possessions from pockets and purses. With unfailing interest, the Clintons examined every one. Through tears, handshakes, and hugs, the relatives of those who died on Flight 800 became convinced the president was feeling their pain.

In Washington, Chairman Hall already knew about the FBI lab finding. He was eager to know what his own people thought of this potential evidence of a criminal act. He relied heavily on the expertise of Dr. Birky, who "never believed it from day one." If Birky was wrong, Hall was more than ready to turn the whole mess over to the FBI.

"He said to me, maybe you guys should take the investigation now," Kallstrom says, remembering how the chairman reacted when he'd learned about the chemical explosive residues.

The Monday after Kallstrom's stunning disclosure at the hangar, Birky had gone to the FBI lab in Washington, a few blocks away from the NTSB office at L'Enfant Plaza. On his return just a few hours later, he went directly to the office of the chairman as promised.

Come on in, Merritt, Hall said, motioning to a chair by the conference table where he was sitting. Birky entered but did not sit down.

I already know what your opinion is on this, Hall offered. I just got off the phone with Louis Freeh.

Oh? Birky was surprised by that. He had left the FBI lab less than fifteen minutes before.

He called to tell me you don't think their tests are conclusive, Hall informed him.

Well, that's right. If they're gonna say that a bomb or a missile took down this airplane, they should have more than they have right now, that's for sure.

Hall waited, confident Birky would continue.

The FBI has positive findings, but these analytical procedures can measure small quantities. These are very small amounts, he said. Birky was just warming up. The overwhelming thing to me was, one, the pieces it came from were not in the right location, and number two, there is no hardware showing high explosive damage, and those two things are pretty fundamental in my book.

Well, Hall said, leaning back in his chair, where do you think these chemicals came from?

There are various ways, the wreckage was put on a navy ship right? Birky asked. And there's lots of ways you can get contamination on an old aircraft.

Look, the chemist continued, I don't think the FBI should put too much weight in what they have here.

And the board, what should we say? Hall asked. What should we tell the public it means?

We can say we found explosives, but it doesn't mean shit.

19

Technology today is capable of protecting the fuel system in its entirety against explosion," Pan Am Captain Eugene Banning said in a speech to fellow pilots in 1968. He had railed on the subject of fuel tank explosions and how to prevent them. He worried that the issue would be clouded by considerations of cost and necessity.

"The cost to provide this protection is not high, not cheap either, but certainly not as expensive as the TV, movies, music and other frills . . . Would the public, if given a choice, choose movies or a good fuel explosion prevention system?"

Banning posed the question rhetorically. Elsewhere, executives with the three American companies building airplanes at the time were answering it.

"You can't smell, see, or taste inerting, and you don't have good-looking women serving it to you," is how one of them explained the industry's aversion to installing inerting systems on airplanes. In a sworn deposition given as part of the lawsuits arising from the crash of TWA Flight 800, the air force's Bill Brookley recounted conversations he had on the subject with Don Nordstrom of Boeing. The bottom line was always the cost.

"He said that they could afford to lose an airplane with a couple of hundred passengers every ten years, and it would still not cost as much . . . as it would to install a nitrogen inerting system on a fleet of aircraft and support it for ten years," Brookley told the lawyers.

In his own deposition in the same suits, Nordstrom denied having said that, or even making such a comparison. Ted Tsue, who worked for Nordstrom at the time, wasn't present for the conversation, though he clearly remembers the attitude reflected in the statement attributed to Nordstrom.

"I think that was the true feeling of Boeing management at that time."

In 1969, with lobbying from Parker Hannifin, the FAA decided to take the liquid nitrogen inerting system out for a test drive. Brookley, serving as a fuel system safety engineer at Wright Patterson Air Force Base, was asked to help the FAA on the project. The unusual request acknowledged the expertise being developed in the fuels hazards branch at Wright Pat.

Brookley worked with the FAA's research and development branch to install the system on a DC-9 owned by the agency. How well it performed under scrutiny would be used to gauge whether the costs were worth the benefits to the commercial airline industry. The DC-9 flew for the last six months of 1971 with detailed records kept on the performance of the system and the plane. At the conclusion the directors agreed it was a reliable method of explosion protection requiring minimal maintenance or attention from the flight crew.

Thomas Horeff, the project director for the FAA, remembers how pleased he was with the work his people had done. "This was the first time that an inerting system was incorporated in a commercial airplane," he said proudly. "This was a major accomplishment."

Even as the FAA was examining the benefits, the need for some kind of protection was made deadly obvious. On November 27, 1970, a fully loaded DC-8 jetliner crashed on takeoff in Anchorage, Alaska. Two hundred and twenty-nine people were aboard, but only 180 survived. The other forty-nine people were unable to escape the airplane before fire triggered an explosion in a still-intact fuel tank. Just seven months later, an Allegheny Airlines plane crashed in New Haven, Connecticut. Only one of the twenty-eight passengers died from the impact. The rest were killed because they were trapped by the fire and fuel tank explosions.

Seventy-seven more tombstones. The National Transportation Safety Board asked the FAA to add fuel system fire safety devices to its airworthiness requirements. In the letter accompanying the recommendation, safety board chairman John H. Reed suggested that the time was right. He applauded the efforts of the FAA so far, writing, "Particularly encouraging is the operation of your DC-9 aircraft with an operationally functional explosion/fire suppression system." The Air Line Pilots Association and the Aviation Consumer Action Project, a Ralph Nader group, joined the plea.

Still, Horeff had his doubts about whether the FAA would mandate fuel tank inerting. "I know, like every other government employee knows, the Office of Management and Budget would not accept any proposal unless the benefits outweigh the expenses. That's standard government policy. Everybody at FAA, NTSB, and NASA knows that."

The FAA proposed requiring explosion protection systems on all commercial jetliners in 1974 and solicited input from the public. It was eleven years after the crash of Pan Am Flight 214. Captain Banning and Bill Brookley were elated. They didn't realize it would take three more years to bring the issue to public debate. It might have taken even longer had the world not been shocked by the deadliest crash in aviation history.

It was a runway collision between two 747s on the small Canary Island of Tenerife on a cloudy March day in 1977.

Visibility was bad. A Pan Am jumbo jet was still on the runway, but the captain of a KLM jetliner could not see it. He began his takeoff roll prematurely. When the departing pilot spotted the Pan Am plane through the fog, he tried to take off over it, but it was too late. In the resulting fires only sixty-one passengers escaped. Five hundred and eighty-three were killed.

Within days of the disaster, Brookley says he got a visit from William "Joe" Sullivan, from the FAA's regulation division. The fact that the proposed rule on fuel tank inerting had languished for three years was obviously on his mind.

"I worked for the Aeronautical Systems Division. Right after the crash in Tenerife, Sullivan flew to Wright Patterson and said he needed presentations. 'We're being pressed hard for a public hearing. The pressure is such that we need to hold it now,'" is Brookley's recollection of Sullivan's reaction.

Three months later, the FAA heard testimony on the proposal. It was fourteen years after Pan Am Flight 214, yet it was as if no time had passed at all.

The manufacturers and airlines claimed, as they had in the sixties, that fuel tank inerting was too heavy, too cumbersome, and too expensive.

Fred Grenich represented Boeing at the hearing. "Our evidence was strong enough, and in my mind we presented overwhelming evidence that liquid nitrogen inerting was not viable for commercial use."

Regulators agreed. After more than a decade of research, and the formation of several government/industry study committees, the FAA concluded inerting was premature. Once again, what the industry dismissed as unworkable, the military jumped in to support. In the months before the FAA hearing, the Directorate of Aerospace Safety issued a report that raved about the liquid nitrogen inerting system on the C-5 military transports. "By saving one C-5, the inerting system more than paid for the cost of modifying the entire fleet," the report concluded.

Still, liquid nitrogen wasn't workable for certain types of military planes. Some aircraft using it needed servicing after each flight. Wright Pat's Bob Clodfelter felt the technology had gone as far as it could go. It was time to try something new.

"We started working on the gas generators, that's the bleed air recycle plan," Clodfelter explained. Unlike the liquid nitrogen system, which required carrying heavy, super-cold, highly pressurized tanks on the aircraft, the onboard inert gas generating system, called OBIGGS, created the nitrogen for the fuel tanks by separating it from engine air.

The FAA showed little more than a passing interest. "We did continue with studies of OBIGGS nitrogen generation. We issued contracts to General Electric and Air Research for ways to produce nitrogen," the FAA's Tom Horeff recalled. But mostly, the studies attracted the attention of the Air Force.

"Those initial studies were expanded on for the Air Force. After that, the FAA phased out of the research and development for inerting," he said.

By 1981, the Air Force was issuing its own contracts to incorporate OBIGGS on military aircraft. Boeing got the job, and one of the engineers assigned to the project was, ironically, Fred Grenich.

Throughout the eighties, Grenich and another Boeing engineer, Charles Anderson, worked on ways to separate air into its two components, oxygen and nitrogen, discarding the oxygen and using the nitrogen to fill the tank's vapor space in a much simpler way than with the liquid nitrogen system.

"We reduced the size and the weight by factors of ten, which was very impressive," said Grenich. Impressive enough for the Air Force to order Boeing's OBIGGS inerting system installed on the large C-17 troop carrier McDonnell Douglas was going to build.

Anderson remembers that the reaction to the Air Force decision at Wright Patterson's labs was muted but very pleased.

"Everybody went between offices and said, 'Hey, did you hear? Great, good news.'" Clodfelter says it felt like a big hurdle had been crossed. "We celebrated. We were happy to get that agreement."

So delighted was Clodfelter with the progress on inerting that he talked it up wherever he went. "I sent information to the FAA and other people and gave papers at different technical meetings.

"I would just tell 'em what we were doing from a technology standpoint," he continued, "what the system looks like, how it works." Clodfelter and Grenich were convinced inerting technology had advanced so far that it was practical for commercial airliners.

"The military decided they were not going to rely on control of ignition sources and each service decided to control the flammability of the fuel itself. Commercial should have," concluded Bob Zalosh, professor of fire protection engineering at Western Polytechnic Institute.

But commercial did not. The NTSB issued no more recommendations, the FAA turned its attention elsewhere, and Boeing's studies as promising as they were remained in the military arena until the crash of TWA Flight 800.

"There was no on again/off again," explained Richard Hill, the FAA's fire safety expert since 1969. "This program was on and then it was off."

The FAA had a new idea for fire safety in the eighties, antimisting kerosene, a fire-suppressing additive to Jet A fuel. The product held great promise for preventing some post-crash fires, but it was useless against fuel tank explosions. In a compelling test in the California desert in 1984, it proved useless against post-crash fires as well.

It was a hot day, the first day of December. Reporters and photographers assembled along with then–Secretary of Transportation Elizabeth Dole to view one of the most dramatic demonstrations of fire prevention research. The FAA along with

NASA was going to crash a radio-controlled four-engine airplane loaded with dummies strapped into seats and show the world how anti-misting kerosene increased survivability.

More than four years of testing and analysis had gone into preparing the Boeing 720 for the crash. The aircraft flew for nine minutes, crashed on a dry lake bed as planned, and promptly and unexpectedly erupted in a tremendous ball of flames, which burned for an hour.

Like fuel tank inerting, anti-misting kerosene was dead.

Fred Grenich had already left his job working on inerting technology for the military and was assigned to Boeing commercial when TWA Flight 800 crashed. He got a phone call from his old friend Ivor Thomas asking him to join an in-house task force looking into what might have caused the 747 to explode. But before the two men hung up, Thomas said something that indicated Boeing already had an idea about what happened.

Dust off your old data on fuel tank inerting, Fred, Grenich remembers Thomas advising him. I think we're going to need it.

20

On the day before the memorial service for Daniel Cremades in Spain, the Suffolk County medical examiner notified his parents that some bone fragments and part of a skull had been identified. Six weeks had passed since the crash, and they had given up on the hope that their fifteen-year-old son's body would ever be found. The service went ahead as scheduled in the community of Palamos on the Mediterranean. Afterward, Ana Vila, Daniel's mother, flew to New York and brought her son's remains back for a private burial the following week.

Such a somber service was not what Vila and Jose Cremades wanted for Daniel's many friends, though. A much more unconventional event was planned in Strasbourg, France, where the family lived. The parents asked the teenagers to organize it. They wanted them to have an event that would express their pain, but remember the good times, too.

For days, dozens of boys and girls filled the Cremadeses' urban apartment, deciding who would speak, going through CDs and selecting the music. Whenever there was anything to be done in school, Daniel had been in the middle of it.

Watching the kids planning the memorial service this way made Vila and Cremades feel that he still was.

"He didn't belong to us, he belonged to his friends and all the people he shared his life with," his father, Jose, said.

On the day of the service, 200 young people and their parents filled the Catholic church in Strasbourg's historic city center. The juxtaposition of teens holding candles while John Lennon and rap music played in the old church was just the kind of tribute his parents wanted for him. At the end the Elton John tune "Daniel" was played.

> *Daniel is traveling tonight on a plane*
> *I can see the red tail lights heading for Spain*
> *Oh and I can see Daniel waving good-bye*
> *God it looks like Daniel, must be the clouds in my eyes*
>
> *Daniel my brother you are older than me*
> *Do you still feel the pain of the scars that won't heal*
> *Your eyes have died but you see more than I*
> *Daniel you're a star in the face of the sky*

At the end of the summer, FAA administrator David Hinson noted on a visit to the hangar that it was "very clear, very early that there was no missile or any bomb for that matter." In his mind, a problem with the plane was the obvious cause four weeks into the investigation, "as soon as the center tank was reconstructed." Of course, news from the FAA that there may be a reasonable explanation for the residues of chemical explosives on the wreckage helped, too, even if it turned out not to be as conclusive as it first seemed.

On September 19, the FAA's bomb technician, Calvin Walbert, forty-five, was out on a New York police boat examining wreckage for the telltale signs of sooting and pitting that might indicate an explosive device went off in the plane. It had been his daily job for two months, but so far he'd seen nothing suspicious. The call he received on the boat from the FAA's on-scene coordinator, Joe Manno, was the most interesting thing that had happened all day.

Manno told Walbert that a search of FAA records showed that in the spring, at St. Louis' Lambert Airport, a chemical-sniffing dog had been tested on N93119. It had taken the FAA nearly a month to find this out, but then again, no one even started looking into the matter until the FBI announced on August 23 that traces of explosives had been detected on the wreckage.

"It sounds like a no-brainer, that it would be something you would automatically check," Walbert explains, "but there were so many things going on it never crossed anybody's mind." Manno also called Jim Kallstrom.

You won't believe it, Mr. Kallstrom, we found records that showed there was dog training on the plane.

You gotta be shitting me, Kallstrom replied. The traces of explosive residue had been like a bad girlfriend—at first tantalizing, later confusing, and ultimately disappointing.

He listened with half an ear as Manno continued to explain. Okay. Fine. Whatever. It is what it is.

There was a news conference that afternoon, and a riot of tabloid headlines the next day. And after that, the matter was pretty much closed, but it continued to bedevil the investigators.

First, the St. Louis police officer who had conducted the test admitted the chemical explosives had leaked while he was hiding them in the passenger cabin for the dog to find. Neither the officer nor the dog went into the cargo hold at the tail of the plane, where the FBI had obtained the second hit for explosives.

"Where the bureau got hits on wreckage, there was no explosive training aids anywhere near that," Walbert said.

It gets more confusing. After conducting tests on the solubility of the chemicals, Irish Flynn, the FAA's associate administrator for civil aviation security, doubted if explosives in either location on the plane would have survived several days or even weeks in the Atlantic.

"To my mind, it's a question of where those traces came from," says Flynn. "The dog doesn't answer the questions."

The most likely sources for the chemical explosives, according to Walbert, were the same sources Birky had ticked off for NTSB chairman Jim Hall weeks earlier: the military ships onto which the wreckage was loaded, the smaller law enforcement vessels transporting the pieces to shore, or the National Guard trucks that carried them to the hangar, none of which were cleaned up to prevent contamination of the wreckage.

The source of the explosive residues remained an open question, but its importance faded in light of the accumulating evidence that a malfunction in the fuel tank triggered the blast. Bob Swaim remembers he was still aboard the Navy vessel *Grasp* when a large, nearly six-foot-high piece of aluminum was pulled aboard the ship. Watching on the deck, he was joined by Boeing engineer Kelvin Deane, who later identified it as a interior wall of the center fuel tank, called span wise beam two.

It was a big flat, structural piece, it had molten aluminum on it, is how Swaim described it to his boss, Al Dickinson, on a secure radio provided to him by the FBI. It seems to indicate fire in the fuel tank, Swaim relayed, explaining his enthusiasm saying, we'd seen burnt pieces, but it was our first in hand.

As provocative as span wise beam two was, members of Deepak Joshi's structures group quickly discovered there was a limit to what they could do with it initially. They knew which wall it was, but not its exact location, "because the left side and the right side are the same," he explained.

It was enough, Joshi said, to get the investigators thinking about a fire in the tank. "You have to focus on something," he told them, "so we started focusing on the center tank."

The initial clue paid off about two weeks later when a large section of the front wall of the tank arrived and showed a series of evenly spaced, penny-sized depressions going from the interior to the exterior side. Once again, Boeing's Kelvin Deane had an explanation.

"He said, 'I know where this might have come from,'" Joshi remembers. Deane told the group that something must have pushed the first interior wall of the tank, called span wise beam three, forward into the front wall. A line of rivet heads running along the top of the beam left the slight impressions in the front of the tank.

Two large fiberglass storage bottles that hold water for passenger use during the flight are secured just outside the tank. When they arrived in the hangar fairly early on, damaged but essentially intact, investigators set them aside. At the time they seemed unimportant. Yet once the center wing tank wreckage started to come together, and the damaged bottles were put in place, it showed once again that an explosion had blasted pieces of the tank forward. The water bottles were directly in the line of fire.

"There were definite marks on the water bottles that they'd been impacted by the top of span wise beam three," Birky explains.

"The front spar showed the same type of damage as span wise beam three folding downwards from the top and hinging at the bottom," he said.

Now there was more confirmation of an explosion in the center tank.

Hank Gray loved risk, but he wouldn't dream of going to a business meeting inappropriately dressed, so for his five-day trip to Paris he'd packed three business suits, dress shirts, a favorite Armani sports coat, some casual clothes, a pair of jeans, and his workout gear. He was less concerned about the other stuff, stashing his leather briefcase with little more than his business papers and a financial calculator.

He carried the most important thing in his head, a plan to pair his company with another, a pricey marriage that he and Kurt Rhein, his partner in the venture, were confident would be profitable. All they had to do now was come up with $80 million to finance it.

The plan was so much a part of the wheeler-dealer personalities of the men who conceived it that when they died on Flight 800, so did the merger. Although no deal is a sure thing until it's done, just the prospect of this one set things in motion that continued to have an effect on Gray's family and employees for years afterward. It was the job of Gray's older son, Hank IV, to handle the consequences of his father's affection for living life on the edge.

The Gray boys, half-brothers on their father's side, were raised by their mothers as their dad went off to marry a third and then a fourth time. "He just didn't relate to children," said his girlfriend, Tara Tomlin. She was equally nervous about his inability to chose a wife and stick with her.

He was a perfectionist, moody and charming, demanding and brilliant, flirtatious and sexy. "I loved him from the first minute I talked to him," said Tomlin.

As his four failed marriages testify, Gray had no fear of making mistakes, but indecision was unforgivable. His longtime friend and coworker Elena Barham remembers Gray used to brag that he could "make the wrong decision, change his mind, and fix the mistake before some people will ever make a decision."

The merger deal between Danielson and Midland Financial Group looked like a sure thing. At $14 a share, it would have given Gray a healthy profit on hundreds of thousands of shares he'd been acquiring in anticipation of the sale.

"He used his entire net worth to purchase more stock," reported his son. He borrowed money to accumulate more, using the value of his shares of Midland as collateral to purchase stocks worth twice as much.

Gray's plan, according to Tomlin, was to make enough money so that after the merger he wouldn't have to work as hard. His value to the company was clear. Danielson would provide incentives for Gray to continue as leader of Midland, but Tomlin said, "When it went through, he thought he could cut back some. He was forty-seven and had visions of not working so hard."

In fact, it was his son Hank who wound up working hard. Following his father's death, he quit his job to work full-time on extricating the estate from all those complicated financial deals.

Within a week of Gray's death, Midland's stock price dropped dramatically. When the value of the collateral deteriorated, the estate received a margin call from Paine Webber. Forty-four percent of Gray's holdings in Midland were sold at $6 a share to satisfy just that creditor, a loss to his estate of $700,000.

"That satisfied the margin call," Gray IV said, "but he also had taken out other loans to buy stock, and we owed that money as well." It wasn't until Midland was purchased by Progressive Insurance Company in 1997 for $9 a share, and Gray IV pitched in money of his own, that the debts of his father's estate were finally paid off.

Midland stockholders took a beating as well, but it was the small company's employees who felt the loss the most. Five of seven offices were closed and 240 people lost their jobs. Midland had been very much a reflection of its bright and quirky boss. Barham summed up the business under Gray's leadership by saying, "It was like a family. It was unique in that respect."

The second time the crash detectives visited TWA's maintenance hangar at JFK, it was an ugly summer Saturday. Once again, they would trek through the tank, but this time an examination of the air-conditioning packs below the tank and a walk through the passenger cabin was part of the tour as well.

Swaim and Birky made the one-and-a-half-hour drive back to the hangar at Calverton together. The rain must have kept would-be beachgoers in their hot city homes, because the Long Island Expressway was remarkably clear as the men headed east.

Birky was trying to understand the explosion that burst the plane apart. Swaim was thinking about what triggered it. Together they turned over some of the things that had made an impression on them, the cleanliness of the tank interior, the soot in the pack bay, the amount of wiring that ran through the tank. They wondered about Boeing's claim to have isolated the tank from all ignition sources. They discussed what might have set fire to the fuel vapors.

They started with the fuel pumps because they had been implicated in other accidents. Other possibilities included hot projectiles, like a broken turbine blade shooting up from one of the air cycle machines below and penetrating the tank; fire from the main landing gear wheel well; or a fuel leak onto one of the hot surfaces below the tank; electrical malfunctions, like an electric power cable melting through the tank's aluminum roof or a short circuit into an electric component inside it. These ideas and more came spilling out of the men as they traveled down the highway.

"We were talking about pumps, things coming down from the cabin, up from the bottom. We were trying to cover all the bases and come up with a list," said Swaim. "When we passed twelve potential ignition sources, we said, 'Wait a minute,'" Swaim explained. "'What else is there?'"

The way to get those answers is by looking carefully at the long and complicated process of certifying an airplane. Manufacturers must show the Federal Aviation Administration that every part of the aircraft meets standards of safety and airworthiness. At Boeing, part of the certification documentation included a failure modes and effects analysis. In 1969 Don Nordstrom was responsible for the certification of the 747's fuel system.

"We assumed failures of various components and then proved that the systems we had designed were redundant sufficiently to prevent explosions..." Among the failures considered by the Boeing engineers, according to Nordstrom, were whether malfunctioning pumps or wiring could cause an explosion in the fuel tank. The dilemma of powered flight is that fuel must be flammable, but flammability creates a hazard.

Boeing based the safety of its fuel system on a design that keeps ignition sources away from the tanks. Boeing's Ivor Thomas explains why.

"Aircraft are divided into different areas of concern. The part class for fuel tanks is a flammable zone. The requirements are spelled out: eliminate ignition sources, keep temperatures below 390 degrees Fahrenheit, and make all components explosion proof."

Boeing accomplished this to the FAA's satisfaction with electrically powered fuel pump motors that are mounted outside the tank and pump housings strong enough to contain an explosion. Where electrical current is necessary for powering the fuel quantity indicating system, voltage is limited to keep the total energy below 0.025 millijoules, about one tenth of the minimum needed to ignite volatile vapors. Bonding the internal components of the tank to the airframe protects against static.

Of course, nothing's perfect, even when hundreds of highly educated and trained engineers put their heads together on a design and hundreds of others review the plans with an eye toward safety. When problems are discovered, the manufacturer and the FAA have an obligation to notify airlines, and, in the FAA's case, order fixes.

The FAA ordered inspection and modification of fuel pump wiring in 747 wing tanks in 1979. Over the years, Boeing issued service bulletins advising its customers to make various changes or inspections for safety.

Airlines must follow FAA orders, but they are not required to heed the Boeing bulletins. In fact, TWA had been sitting on one to inspect fuel pump wiring for a year before the crash of Flight 800, even though the suspected problem was a fire hazard. TWA told the safety board it was waiting for the FAA to order the work, which it did in March 1997.

These facts were surely in the minds of Flight 800's investigators on August 25, when members of the fire and explosion group carefully carried the two center wing tank pumps retrieved from the ocean to a NASA laboratory in Huntsville, Alabama. They wanted to bring the third pump, called the scavenge pump, because it had the most provocative maintenance history, but that pump was never found.

NTSB records inspector Debra Eckrote reported that in the two years leading up to the crash, the scavenge pump was in need of service nine times and was replaced twice.

Still, it was considered unlikely that any of the pumps contributed to the explosion because they were not running at the time. One of the examinations conducted at NASA showed that the cockpit switches controlling the pumps were in the "off" position.

When all the evaluations and test explosions on fuel pumps were done, there was no evidence found of an overheated motor or an internal fire in the fuel pumps. "If you had an electrical malfunction, you might see the evidence of an electrical arc that would be dark or burned spots on the insulation," said Stan Bluhm, director of mechanical engineering at the time for Hydro-Aire, manufacturer of the fuel pumps. "We found none of that on any of the pumps we inspected, though not all of the electrical connectors were there. Those that were showed no signs of electrical arcing or overheating."

The next step for investigators was to look at the ways an electrical charge could build up in the tank, a difficult task, because static electricity leaves no trail, unlike mechanical failures. "Static is a ghost" is how aviation researcher Michael Barr describes it.

Swaim's examination of N93119 and several other 747s of the same age showed unbonded metal components in the tank. If a static charge collected on that metal, it could have caused a spark. Boeing tries to prevent this by using bonding straps to connect the metal in the tank to the structure of the plane. It's similar to how electricity is grounded in a home.

But what if one of the bonding straps, some of which are quite fine, had broken? The question took Bob Swaim to Joseph T. Leonard, a civilian scientist formerly with the Naval Research Lab and a specialist in the field of electrostatic charging of fuel. Together, the two men traveled from Washington to the hangar at Calverton. Leonard, who had investigated several accidents in which static ignited fuel tanks, walked around the hangar with Swaim. The contrast between the enormous reconstruction going on in the hangar and the fragile, fleeting theory of ignition he was bringing to the investigation was not lost on either man.

Walk across a carpeted room and touch a doorknob, Leonard was telling Swaim, you get a shock, that's about ten millijoules. When it really smarts, that's about twenty. One quarter of a millijoule, the energy needed to ignite the fuel tank being pieced together in the very hangar where they stood, that, he explained, you wouldn't even feel.

The theory that static had accumulated on an unbonded metal surface in the center tank of the plane relied on the assumption that a leak had developed in a pipe that runs through the center tank carrying fuel from wing tanks on one side of the plane to engines on the other. Investigators had reason to think fuel sprayed into the center tank because the 747 left Kennedy with fifty gallons of fuel, but at the time of the crash the gauge registered 100 gallons.

Fuel spray can generate static, though to Leonard the theory "didn't sound like a typical static scenario. I thought it was unlikely but I thought it was worth investigating."

To test the possibility, Leonard went to the labs at Wright Patterson. He sprayed various types of kerosene onto the two metal components investigators thought might have been unbonded on N93119, a small clamp and a coupler on the cross-feed manifold. He could get static to build up, 500 volts on the coupling, 650 on the clamp. At that point, the electricity always jumped off onto the grounded side of the metal. It happened repeatedly, before the metal could collect enough voltage to create a spark.

After three months of testing using conditions likely to have been on Flight 800, Leonard concluded, "We just can't charge it highly enough."

It was good news for the flying public. This kind of static was an unlikely source of ignition even when tank components became unbonded. "We couldn't do it that way and I doubt if it could be done," said Leonard.

But it was bad news for investigators. They were back to square one.

Reporter Tonice Sgrignoli spent the night of July 22 in jail in Queens. She faced five felonies including criminal trespass, criminal impersonation, and possession of stolen property, that being the small metal button she wore each day giving her access to the Ramada. She was strip-searched and put into a holding cell with a man who'd obviously had too much to drink. He was sleeping it off, and for that she was thankful.

The contrast between having been coddled and cared for in the protective bubble of the Ramada and her treatment at the hands of the New York police didn't really surprise her. Earlier in the day at the memorial service, she herself had shuddered at the baying press corps.

"You saw this faceless horde, camera lenses instead of faces, and they were always grasping to get at you. And I couldn't relate to my peers," she said.

"I didn't feel like a reporter very much because the press was the enemy." Sgrignoli knew the police who now had her in their custody probably felt just as she had, but now she was the enemy, too. "They treated me like scum. They finally had a member of the press they could take it out on."

It took months for Sgrignoli's case to make its way through state court in New York. The newspaper paid a fine on her behalf, and she was ordered to write a letter of apology to the woman she'd befriended, submitting it to the court for approval.

Sgrignoli very much wanted to communicate with the woman. She just wished she had been able to do it freely, and without restriction. When she returned to the *New York Post* the morning after her arrest, she was greeted with a standing ovation from her peers. Days earlier she wanted to distinguish herself. Now she saw their response as dismissive of the questionable nature of what she'd done. Three days at the Ramada had taken her somewhere many reporters have never been, to the intersection of journalism and compassion.

2 1

Bernie Loeb, the man to whom the crash detectives reported, had a clear idea of what still needed to be done. Rebuilding the airplane was not on the list.

As his boss, NTSB chairman Jim Hall, was assuring Jim Kallstrom that rebuilding would soon start, Loeb was doing what's come to be known at the safety board as the "Bernie's Bernie" act, stalling and delaying, refusing to go along with anything he objects to. In his opinion, rebuilding the plane was a waste of time, an attitude shared by Merritt Birky.

"I agreed with Mr. Kallstrom that we had to have the whole thing, the whole plane reconstructed," Hall said, but it had to be balanced with his "great deal of respect for Bernie and Merritt Birky."

At the FBI there was no difference of opinion about rebuilding the plane. Pan Am 103 had been reconstructed. There was no reason not to do the same for TWA 800, if for no other reason than to rule things out definitively, including the irrepressible missile scenario.

"We as a group," Kallstrom said, "thought we had to look at how the holes lined up and ask, 'Did anything penetrate into the fuel tank?'"

With both sides intractable, the dispute ended up at the White House for resolution.

"Suck it up, and we'll figure it out," cabinet secretary Kitty Higgins would have liked to have said, when she was called once again to step in between Hall and Kallstrom's conflicting agendas. Instead, she arranged for a meeting with White House chief of staff Leon Panetta in November.

"The tension was, they each thought they could define the investigation and they just couldn't," Higgins said. "It wasn't theirs to call."

For Kallstrom, the argument for building the mockup was simple. How could I ever answer the question, "What happened?" he asked Panetta. How could I say we're sure about anything?

We're sure, was Loeb's reply. They had all the evidence needed to rule out a bomb or a missile.

Bernie, you're not in charge of determining if this was a bomb or a missile, Kallstrom said. When the investigation is done, you can say what it is.

When the men had made their points, Panetta looked at them and told them to build the mockup. Loeb might have been disappointed that his opinion was not strong enough to carry the day, but his boss later said that Loeb's arguments had served their purpose.

"Let me tell you the reason that I let Bernie speak," Hall recounted of the meeting. "It was because of the amount of money involved. I wanted the White House to know what they were getting into."

Loeb remained unconvinced of the FBI's contribution to the investigation. "The FBI was clueless," he said later. "They never have been involved in plane crashes. They act viscerally on what they see."

Kallstrom was equally dismissive of Loeb. "Someone's personal opinion doesn't mean shit," he said.

On February 8, 1997, 163 people who lost someone on Flight 800 came to the hangar at Calverton to view the $500,000 reconstruction Panetta had ordered. In the years to come, the project the safety board didn't want to build would, ironically, be put to use as a teaching tool for NTSB aviation accident investigators.

The squabble over the mockup had been about whether the money and effort were well spent, whether, once it was done, the rebuilt fuselage would even be elucidating. What both agency heads had not discussed was the impact their creation would have on those most affected by the disaster.

By February, the ninety-foot center section of the plane was substantially complete. Tiny shreds of the plane's lower skin were painstakingly reattached and the larger sheets of metal curled out away from the structure uncontrollably in dozens of areas. Tremendous tendrils of cables and wires drooped toward the floor. The right side of the fuselage over the fuel tank was coated in a film of ashy soot. The destructive forces at work were powerfully evident.

In the smaller cabin reconstruction hangar, the passenger seats had been put together ten across and mounted as best they could into positions roughly as they had been on the 747.

It was cold and snowy when the families came to see the work. They toured in small groups, accompanied by people from the FBI and NTSB. Bob Francis, who had forged relationships with several of the family members, remembered that the tour was "tough stuff for everybody." Some people wanted to place flowers on the seats. Because the wreckage was taped off, Francis had to do it for them.

Stephanie Maranto's brother, Jamie Hurd, had been sitting in the right-hand window seat of row thirty-three. After the visit, she choked back tears as she told reporters about the experience.

"When you look at the empty seats that are charred and busted up and mangled, you immediately run through your mind, these seats were full at one time. They were full of life and people and children and husbands and wives, and it's upsetting to see them sitting there like that, completely empty and just broken to pieces. It's a very quiet, cold, and eerie feeling."

Jose Cremades, the father of fifteen-year-old Daniel, had flown over from France for the event, which he described as "a pilgrimage." Joseph Lychner added, "I was hoping for a clear indication that my family died instantly. Unfortunately, because of where they were sitting on the plane, that was not clear at all to me."

As Cremades and Lychner stood talking with reporters, they dismissed the possibility of the plane having been brought down by a criminal act.

"Was there a design defect?" Cremades asked. The Spaniard, an official with the European Union, had been briefed regularly over the previous months by safety board officials, so he was fully aware investigators still did not know what caused the plane's center tank to explode. Yet he already understood the position the board would take in the future when he added, "These planes fly with a design that allows a single failure to cause a crash. The source of ignition is unimportant."

Disarmingly frank, the NTSB's Peter Goelz describes Bernie Loeb as "both a sword and a shield," acknowledging that the aviation safety chief is used to advance the will of the chairman while deflecting any criticism.

It explains why it was Loeb who began to pull the investigation back into the safety arena. He started in the early fall, telling industry publications and major newspapers that the crash of Flight 800 was probably "a mechanical issue." Then producers for CBS's *60 Minutes* were offered an on-camera interview with Loeb for a program on the same theme.

Jim Kallstrom was in a fury when he heard about the planned report. He called up *60 Minutes* and asked that his position be included in the broadcast, which it was. At the same time, he complained to Hall that the agency's speculation jeopardized the criminal investigation.

Hall was having none of it. He felt Kallstrom was responsible for some of the most damaging speculation by saying in the early days that he was going to find a "Eureka" piece, that he was going to find the "cowards responsible." The FBI sure had benefited from the bluster. The Aviation Security and Terrorism Act, providing millions of dollars to the FBI and expanding antiterrorism measures, had passed Congress with help from Kallstrom's testimony.

"There wasn't adequate attention on what we felt was the strongest possibility, which was an explosion in the center tank," Hall said, justifying the decision to separate from the party line that all three theories, a bomb, a missile, and a mechanical malfunction, were being given equal weight.

Loeb was laying the groundwork for emergency safety recommendations about to be proposed by the safety board.

Suggestions that Flight 800 had self-destructed prompted Boeing officials to begin their own media campaign. Senior engineers called reporters around the country and insisted "chances are zero of that happening."

Boeing engineers had claimed from the start there was not enough punch in a fuel-air explosion to take apart a 747. Literally speaking, they were right. The explosion, the equivalent of ninety pounds of TNT, accelerated as it squeezed through chamber after chamber of the tank, bursting through the front and causing a major support member of the plane to bend down. After that, structural failure was to blame for the crash. It was a distinction without a difference. After all, the plane was down, the people aboard, dead. The next step was to figure out what went wrong.

A little more than five weeks after the crash, the manufacturer outfitted a leased 747-100 with temperature gauges in the tank, let it sit with the air packs running for several hours in the Mojave desert, and then flew it to 13,700 feet.

Dennis Rodrigues, the Boeing investigator at the hangar, says he told the safety board of the planned flight. Notes taken three days prior back up his claim. NTSB investigators, however, say they didn't learn about the flight test until after the fact, from gossip on the hangar floor. They deduced that Boeing was working on its own—gathering but not sharing information relevant to the probe. Chairman Hall was livid.

"Boeing was busy lobbying their own cause," was how he characterized it. "I just told them if I caught 'em again going outside to do anything, that I was tossing them out of the investigation."

The warning was noted by Boeing, and two days later, the information from the test flight was brought to Washington along with an apology and an explanation.

Boeing's Ivor Thomas claimed the opportunity to do the test had come along quickly. "Better to get the data then and argue about whether we'd done the right thing, than not have the data."

The arguing between Boeing and the board was overshadowed by test results showing the tank could get hot, hot enough to have an explosive brew in the tank before the plane ever left the ground.

"It explained the accident," the NTSB's explosion expert Birky said.

"You can have a flammable mix in the tank. If there's an ignition source, it will blow it up. It will blow the plane out of the sky."

The day before Labor Day is not a big shopping day at Macy's in New York City, so few people saw Peter Michael Santora as he was escorted out of the famous Thirty-fourth Street store in handcuffs. Had anyone realized why he was being arrested, it would have drawn attention, as all events related to the mysterious crash of Flight 800 seemed to do.

The mystery of Santora's arrest for grand larceny was how he thought he could get away with it. He was a fifty-year-old white man using the driver's license of an Asian woman to open a credit card account in the name of someone who'd been dead for nearly two months.

He'd swapped Judith Yee's photo for his own on her driver's license, but the application still raised eyebrows. "The associate that took the application obviously realized that something was not on the up and up," Ronnie Taffert, a Macy's spokeswoman, said.

At the police station, police discovered that Santora was carrying eight of Yee's credit cards. Ronald Yee was notified, along with the attorney representing her estate. But it was too late. By their estimates, Santora had already stolen in excess of $13,000 and some of Yee's personal possessions.

How he accomplished the theft was surprisingly easy. He'd volunteered to help Ron Yee close out some of the details of his sister's life.

"We thought he was Helpful Harry, helping people get things done and facilitating things," explained the lawyer for the estate, Gerald Dorfman.

That's one reason, Dorfman says, no one notified her creditors, including her bank, of her death. "There was no need," he explained. "He was a trusted friend."

Dorfman says that from July 18 until his arrest six weeks later, Santora made purchases on her card and removed the billing statements when they arrived in the mail. His actions make it appear he was planning to keep it up for a while, because he paid some of the bills using money from Yee's bank account. He also used her ATM card.

Prompted by Santora's arrest in September, Dorfman closed Yee's accounts and stopped her mail delivery. Friends of Yee's assembled a list of what they believe was stolen, including a personal computer and several pieces of Yee's collection of expensive paperweights. For some reason, charges filed against

Santora in September were dismissed. No record exists of his arrest beyond a newspaper account. Yee's computer and documents of hers that were found in Santora's apartment were taken to the Manhattan district attorney's office. They were never returned, according to Dorfman.

"He's a con man," Dorfman said of Santora. "He wormed his way in with Ron."

Speculating on why the district attorney let the case go, Dorfman suggested Ron Yee might have been embarrassed. "He was duped," Dorfman said. "People who are conned are reluctant to go to the police."

Barbara Thompson, a spokeswoman for the district attorney's office, said there are no charges on file against Santora in connection with Judith Yee. At the time of his arrest in September, police officials were telling the *New York Daily News* that he'd been arrested before for illegal credit card use and he had not paid a $4,000 restitution in that earlier case.

On December 13, 1996, five months after the crash, the NTSB sent recommendations to the FAA asking regulators to change the rules so that commercial aircraft would not fly with explosive fuel-air vapors in the tanks. For the third time in three decades, the accident investigators were urging the FAA to do something about the hazard.

It wasn't just a matter of whether Boeing and other airplane manufacturers were as capable as they claimed of isolating ignition sources, though there was a great deal of doubt about that. The world was a different place. The board reminded the FAA that a well-placed bomb had brought down an Avianca Airlines 727 in Colombia in 1989. It was a small explosive, but it ignited volatile fumes in the plane's center wing tank. Who could control those ignition sources?

Considering how long it takes for even the simplest of actions to wend through government bureaucracy, the Flight

800 proposals got a relatively quick response, though not the action requested. Studies were ordered, as were inspections of pumps, wiring, and fuel tanks.

The FBI's response, on the other hand, was surprisingly hostile. Kallstrom called the safety board's actions premature and argued that if the crash turned out to be a crime, the safety board had just handed the perpetrator a defense.

At the hangar, FBI special agent Dennis Smith reported he felt blindsided. In a note to his superior, Smith wrote that the agency wanted to "turn off the lights at the Calverton facility, dispose of the wreckage, raise havoc within the commercial aircraft community, and send everyone home."

Yet the safety board's opinion was being ratified by the Bureau of Alcohol, Tobacco and Firearms, which drew up its own report one month later, essentially confirming a malfunction on the aircraft as the cause of the crash. Andrew Vita, the ATF assistant director, scribbled a memo to his boss: "We have," he wrote, "evidence of possible design flaws" in Boeing airplanes.

Kallstrom wasn't pleased that the NTSB was drawing conclusions, but he could do little about it. After all, the board was responsible for the investigation of aircraft accidents. The ATF was another matter. About them he complained, "You can't stand up and make judgments without finishing a process. We still had hundreds of things we hadn't done yet."

It was another public eruption of interagency battles that had been going on since the start and would continue long after the FBI turned the hangar over to the NTSB. "I don't know if it's the personality of the chairman or if it's testosterone," cabinet secretary Higgins wondered once in exasperation.

Chairman Hall was determined that in future crash investigations the role of the safety board would be undisputed. Two and a half years after the crash, he had the agency's chief counsel, Dan Campbell, working with both the ATF and the FBI on agreements that said as much. The ATF was cooperating, but the FBI balked, until Iowa Republican Charles Grassley, an eager senator with a long-standing antipathy for the FBI, stepped into the mix. At a congressional hearing in 1999, notable for the fact that Grassley was the only senator present, he criticized the FBI for its actions in the hangar and its response to the ATF report. Jim Kallstrom had already retired from the FBI, taking a job as head of security for the MBNA Bank of America in Delaware. Senator Grassley was castigating Kallstrom and the bureau for having "commandeered" the probe and "hindered the investigation."

Loeb complained to a Grassley staffer that Kallstrom whipped up a frenzy over the potential for criminal activity and then kept the FBI involved long after it was clear the crash was not a crime. Since Kallstrom was no longer a government employee it was left to Lewis Shiliro, a soft-spoken, generally unflappable New Yorker who replaced Kallstrom, to speak up in defense of the agency. His moderate comments at the hearing were a contrast to Kallstrom's outrage, but the point he made was the same.

"I don't see what we would have done different. We did what the American people expected us to do."

"The whole thing is ridiculous," Kallstrom scoffed from his office in downtown Wilmington. "Senator Grassley in one broad stroke tried to discredit a lot of people who worked their butts off."

Two months later the FBI and the NTSB were back at work negotiating the agreement Hall had been seeking, though events soon eclipsed the talk. An EgyptAir 767 with 217 people aboard had flown into the Atlantic sixty miles south of Nantucket Island, Massachusetts, on October 31, 1999, and the two agencies were together again. It was obvious the leaders had learned from past mistakes. Like Flight 800, the cause of the EgyptAir crash was not immediately apparent. The FBI sent agents to the scene, including TWA veterans Schiliro and special agent in charge Ken

Maxwell, from the New York office. More agents were assigned from Boston and Washington, D.C.

The NTSB's involvement came at the request of the Egyptians. Chairman Hall quickly decided he would go to the scene, replacing the board member on call, John Hammerschmidt. This is a carbon copy of TWA, he told his aide Peter Goelz, explaining his action. There are multijurisdictional issues here. Goelz affirmed the chairman's decision, saying, You gotta go.

From the very first news conferences the afternoon of the crash, Hall was the official spokesman for the investigation, sharing the cameras with the FBI's Barry Mawn only once as the lawman announced there was no evidence of a criminal act. When he was concerned that the FBI was leaking information, Hall went right to FBI Director Freeh to complain.

"There was a certain amount of elbowing at the beginning," Goelz said, "especially over who was going to do the press conferences." Both Hall and Schiliro praised the EgyptAir investigation as a great improvement over the chaos following TWA Flight 800.

"We both learned a lot about our cultures," Schiliro told a reporter. Safety board investigators were pleased that FBI agents kept them up to date on their activities. "They are going about their work, coordinating and communicating with us," Hall announced with satisfaction.

A little more than two weeks after the crash, the safety board, suspecting the EgyptAir crash to be a criminal act rather than a mechanical problem, began the process of turning the probe over to law enforcement. The Egyptian government protested the move, suggesting that the speculation that a copilot may have purposefully sent the wide-body into its death plunge was premature. The aviation experts stayed in, throwing the question of which agency would be the lead back into murky uncertainty.

22

While Pierre Salinger kept the mainstream press hopping in New York, James Sanders, a considerably less known journalist, was meeting with Terrell Stacey, a 747 captain assigned to the probe as a representative of TWA.

As contacts go, Stacey was a good one. He'd flown the 747 for years and had brought N93119 in from Paris the day before. He interviewed eyewitnesses, reviewed the cockpit voice recorder, and helped in the reassembly of the interior of the plane as part of the official investigation. Stacey found many things about the probe distressing, particularly the atmosphere of mistrust and secrecy at the hangar.

The FBI's Kallstrom doesn't dispute that things were often tense at the hangar. "There was grousing, 'They're secretive and not telling us shit and bringing stuff into rooms.' We were doing all of that," Kallstrom said, explaining, "They weren't used to being in the middle of a criminal investigation with all the rules and laws that affect that." Kallstrom tried to make it clear to the parties, "It's not that you're not patriots, it's not that you're not hardworking people. We're forbidden to tell you."

Interestingly, Stacey was not among the grousers. Senior NTSB and FBI officials don't remember him lodging a single complaint while he worked at the hangar.

"He was a real gentleman, very quiet. He wouldn't say a word with a mouthful," is how the FBI's Ken Maxwell described him.

Still, when Liz Sanders, a flight attendant instructor Stacey had known for five years, called him in early November to ask if he would talk to her husband, who wanted to write an exposé about the crash, Stacey agreed.

Sanders, a former small-town police officer, had already written a book on the plight of Vietnam War–era prisoners of war. Stacey had read it and was impressed with the author's contacts and resourcefulness. The two men talked on the phone and got together shortly after that. Stacey gave Sanders documents from the investigation and ultimately agreed to remove some samples of a thin red foamy material that was on the back frame of some seats located over the center tank.

Stacey was not eager to remove this physical evidence from the hangar, having been warned repeatedly along with every other person working there that nothing was to be taken from the premises.

"There was a heavy burden with the investigation, frustration with the investigation, the lack of sharing the information by the NTSB, and, of course, the FBI," Stacey explained when asked why he finally agreed to supply the sample to Sanders. "I thought this would be a means of me obtaining some more information, more analysis to find out the cause of the accident."

"Terry is a really nice guy," said Kevin Darcy, a crash investigator for Boeing for eight years at the time Flight 800 went down. Darcy's was a common sentiment. "I felt bad when I found out he was involved in that. He really just felt there was something going on that was being covered up," was Darcy's thinking.

Stacey may have been motivated to scrape the small swatch of red foam because the FBI had already done the same thing. "We boxed it up and shipped it down to the lab," Maxwell said, but no one ever reported back on what the material might be.

Stacey worried that investigators were hiding something. In fact, the sample had been forgotten about in New York and ignored at the lab in Washington. It wasn't until Sanders's claim months later that the residue was missile fuel that FBI honchos looked in on the progress of the sample at the lab.

"I have a distinct recollection of a certain phone call from Kallstrom that was a catalyst to having that evidence handled more expeditiously," Maxwell recounted. Prompted to get the results, the FBI and the NTSB concluded the foam was glue used to attach fabric to the seat frames. It was a logical explanation considering how uniformly the material was spread along the inner frames of some seats.

Still, suspicions that government investigators were somehow up to no good reverberated in the hangar. Linda Kunz, another TWA employee who worked with Stacey, was convinced that NTSB people were undoing work, moving tags that identified where various pieces of wreckage had been found. She snapped photographs and sent them to corporate executives in violation of the FBI rules at the hangar and very nearly got TWA kicked off the investigation.

"But the very idea of a conspiracy, that there was something going on, that people in the hangar knew about it and were keeping it quiet ... that seemed to be ridiculous," Darcy insisted. "Everybody knew what was going on in the hangar; you couldn't keep anything quiet in there."

On the contrary, Stacey's involvement with Sanders was a well-kept secret for nearly a year. Since Sanders was a journalist, law enforcement had to get clearance from the Department of Justice to obtain phone records and track down his source inside the hangar. In June 1997, FBI agents knocked on the door of Terry Stacey's New Jersey home and questioned him about his relationship with Jim and Liz Sanders.

By that time, Sanders had already concluded that a weapons exercise involving the Navy caused the crash, and that the red foam was residue from missile fuel. In Sanders's scenario, the

106th Air Rescue Wing was monitoring war games that resulted in the accidental shooting down of Flight 800. Major Frederick "Fritz" Meyer is specifically named as a participant.

"I picked up his book and asked if we could sue this guy," was Meyer's first reaction to the charge. "I wasn't very pleased with his opinion that I was there directing traffic." Since Meyer is also convinced the government is engaged in a cover-up, his opinion of Sanders softened to the point at which he began to wonder if the National Guard had played a role.

"He had to make some assumptions," Meyer rationalizes, speaking of Saunders. "As far as I know, maybe my squadron had some involvement I knew nothing about. I was not scheduled to fly that night, I was called that afternoon to substitute for someone else."

In the course of pursuing his theory, Sanders came to know two more established journalists, David Hendrix, a writer for the *Press-Enterprise* in Riverside, California, and Christina Borjesson, a freelance producer for CBS Reports. Both decided to pursue the story Sanders was working on, Borjesson going so far as to accept from Sanders a sample of the foam Stacey had stolen for him. It was her bad luck that Sanders left the Fed Ex receipt addressed to her outside his Virginia home the day FBI agents came to question him.

When the foam sample arrived at the network's New York office, Borjesson brought it into the office of her boss, Linda Mason, who promptly called up the network's lawyers. "It was a piece of stolen material from an ongoing investigation," Mason said, "so we gave it back." CBS did not report Sanders's allegations, unlike the *Press-Enterprise,* which timed a series of articles to the publication of his book.

That the journalists had much information in common is not surprising. They were all staying in touch with Kelly O'Meara, the administrative assistant to New York Congressman Michael Forbes. O'Meara was corresponding with the Pentagon, the Coast Guard, and TWA. She received piles of data and had the time and inclination to share it.

"Because I worked for a member of Congress, I could get official responses. The official responses I was getting were not what I was hearing out of Dan Rather's mouth. The media wasn't reporting what I was getting. This was frustrating."

It was equally frustrating for the FBI's Kallstrom. He sent agents to brief O'Meara, as did the Navy, even though he was pretty sure she was actively engaged in promoting a theory that the government was involved in a cover-up.

"I was aware from people around the investigation that Forbes's office was part of this whole conspiracy thing to some degree," Kallstrom said. "A lot of people were concerned and puzzled by what his office was doing. I didn't know how much he was doing and how much was happening by some strong person with a lot of leeway in his office."

Whatever margin of freedom O'Meara had, she used up the following summer when she arranged for a tour of the reconstructed airplane at the hangar in Calverton. Maxwell agreed to host O'Meara and Diana Weir, Forbes's chief of staff. What he did not know until halfway through the visit is that the third woman accompanying O'Meara and Weir was Borjesson, who was no longer working for CBS.

It's hard to say who was angrier, Maxwell or Kallstrom, but it was Kallstrom who called Congressman Forbes.

"I was furious. Here we were trying to cooperate with the congressional people and one of the staff members would bring someone from a news organization into the hangar?"

Kallstrom had his spokesman Joe Valiquette call O'Meara first.

"He was trying to intimidate me," O'Meara recalls.

Your boss is gonna be getting a call from Kallstrom today, Valiquette said. You brought that woman into the hangar.

No one argued that Borjesson had been introduced at the hangar as a friend of the women, and O'Meara reminded Valiquette of that.

I told you who she was, she argued. She gave you her passport. You let her in there.

She's hooked up with Sanders who wrote that conspiracy book, Valiquette insisted.

O'Meara didn't say anything to that. Sanders was sitting in her office.

The conversation was much different when Kallstrom called Forbes.

"I'm sorry," Kallstrom remembers Forbes saying without hesitating. "It shouldn't have happened. I didn't know it was happening and it will never happen again."

Having worked himself into a state over the incident, however, Kallstrom was not quieted by the apologies.

Look, Congressman, if you have doubts about what we're doing up here, we'll be happy to show you what's going on. But were not going to be hoodwinked or play some stupid game with you.

Whether O'Meara quit or was fired depends on who's telling the story. But she no longer worked for Forbes one month later, and by the spring of '98, she and Borjesson were working for movie producer Oliver Stone. They'd convinced him the investigation of Flight 800 was worth another look.

When the 747 was first purchased by Pan Am, the airline's executives thought they'd bought a big, sturdy airplane that would last them for the next ten years, a interim airplane to get them from the jet age to the supersonic age.

As Boeing was putting the finishing touches on the 747 in 1969, it was just beginning work on an American SST with a ninety percent subsidy from the federal government. When the 2707 became a reality, those capacious jumbo jets would be converted to spend the second half of their lives as freighters.

It wasn't long before America woke up from its big dreams of fast flight to the realization that the 747 had performed as promised, flying more people farther and cheaper. By contrast, the Russian and European supersonic planes already built had limited seating capacity and massive fuel consumption. In 1971, Congress voted against continuing to fund the program.

The American SST was dead. Long live the 747. And live long it did. In 1999, more than a quarter of the 747s flying were more than twenty years old.

In the aviation industry, folks are apt to say that if people took care of their cars the way airlines maintain their airplanes, every-

one would be driving 1964 Chevrolets. But the seeming invulner-
ability of jetliners to the ravages of age was thrown into doubt
when Aloha Airlines 737 lost eighteen feet of its roof in 1988.

The FAA responded by coming up with an aging aircraft program requiring special maintenance and inspections on the
structure of older airplanes. TWA's entire fleet of 747s, including the one that flew as Flight 800, were inspected more fre-
quently and more substantially under the new schedules created to spot age-related structural fatigue and corrosion. Still,
N93119 was no showpiece '64 Chevy. Some of the electrical
wiring providing power to the most basic and the most complex functions of the jetliner had not been examined since
installation at the factory twenty-five years earlier. This was not
an act of omission by TWA maintenance. At the time of the
crash, there was no standard inspection program to determine
if age was affecting airplane wiring or other systems.

According to Kenneth Craycraft, a 747 maintenance supervisor for TWA for the entire twenty-eight years the airline flew
the plane, no further inspection was needed.

"When it's opened up, the wiring is quite visible," Craycraft
said, offering that during the heavy maintenance checks a 747
undergoes on average every four years, much of the wire is
looked at.

Ed Block is a gadfly, a persistent, proselytical man of great
earnestness and energy, whose specialty is aircraft wiring. From
1974 until 1984, Block was in charge of purchasing wire for
military planes.

"A lot of people think of wire as nuts and bolts, like an inani-
mate object," Block told a reporter after the crash of Flight 800.
"It isn't. It's more like your nervous system or your circulatory
system."

While working at the Defense Industrial Supply Center in
Philadelphia, he'd learned a lot about wire, both the conduc-

tor—the metal that carries the current—and the insulation—the nonconductive material that surrounds it, keeping the electricity contained.

In 1978, Block's office was instructed to get rid of a wiring product called Poly-X, manufactured by Raychem. Further, its use was prohibited in certain critical areas on the Navy F-14 Tomcat. It seemed the insulation on the wire was breaking down, allowing the electricity to enter areas where it could cause heat or fire or send unwanted electric impulses to flight controls.

Block thought commercial aviation should know about problems with Poly-X, because it was in use on many passenger jets, the 747 and the DC-10 among them, which is how his Paul Revere act with aviation authorities began.

In 1992, Block wrote a letter to the General Accounting Office raising the issue of aircraft wiring insulation and asking what experts the FAA and NTSB had on staff to analyze the role of wiring in aviation accidents. Letters went back and forth, some more responsive than others, but the letter he sent one month before the explosion aboard Flight 800 to Tom McSweeny, FAA director of aircraft certification, seemed almost prescient.

Block wrote, "Just as it took an Aloha Airlines 737 to flip its lid to focus attention on metal fatigue, similarly wire and cable has virtually been ignored."

The response from McSweeny didn't arrive until November, four months after the crash of Flight 800. In it McSweeny concluded, "the FAA has no plans to update or make changes to its wire performance standards. This is partly due to the fact that the wire industry is providing outstanding products to the airframers . . ."

Block doesn't give up easily. On the subject of aircraft wiring, Block felt nothing less than a calling from God to stay on the issue. Vice President Al Gore may have been the one to give Block's arguments the biggest boost.

In the feverish first days following the Flight 800 disaster, President Clinton asked Gore to lead a commission to investigate ways to make commercial aviation safer. When wiring began to eclipse terrorism as the likely cause, the commission urged the FAA to look again at the deteriorating effects of age, this time on electrical wiring, connectors, and electromechanical systems. The FAA's McSweeny was beginning to realize there was some value in what Block had been saying.

Doing as instructed, the FAA issued a report a year later confirming fears that wire inspections were "too general." The report explained there was "no systematic process to identify and address potential catastrophic failures caused by electrical faults of wiring systems aside from accident investigation associated activities." That means, in essence, there was no plan to find out if an electrical problem could cause a crash, until it did.

From these discoveries the FAA took the next step, forming an Aging Systems Task Force to look further into the problems and suggest solutions. A consumer group called the Air Disaster Alliance insisted that Ed Block be given a seat on the committee. His decade as an outsider was over.

"He's frustrating to deal with," McSweeny conceded of Block, putting him in the same category as consumer advocate Ralph Nader. "But we are eagerly working to get his input and everybody else's."

In the first year of the investigation, the crash detectives had a basket full of ideas about what lit the tank on TWA Flight 800. The NTSB's Bob Swaim had been up and down on fuel pumps; hopeful, then disappointed, about static ignition. As theory after theory fell for lack of evidence, interest in the wiring aboard N93119 intensified.

One area that seemed worth closer examination was Boeing's practice of running wires through the airplane in thick, wrist-sized bundles. As it runs its course from the electronics bay beneath the cockpit to the center tank, the low-voltage FQIS wiring, which powers the fuel quantity indicators, is routed with as many as 400 wires carrying loads of from five to 192 volts. Some of that wiring was the Poly-X wiring that worried Block.

"The FQIS wire is co-routed throughout the airplane with the wiring of all other airplane systems. There's no segregation at all. So we're literally talking about hundreds of other wires that are bundled with the FQIS wiring," explained Swaim.

If one of those higher-voltage wires short-circuited into the low-voltage FQIS wire, could the higher current be carried into the tank? Investigators had reason to be concerned. On the TWA airplane, Swaim's team found numerous cracks in the insulation of the Poly-X wires, though no proof that electricity had jumped over to the FQIS wires. They knew, though, that it was a possibility.

Ted Tsue spent thirty-eight years at Boeing before retiring in May 1996. He was a safety analyst on the fuel system of the original 747 design. Tsue explained that engineers working on fuel tank wiring had to "make sure the gauging system is fully isolated," so that only a level of energy less than needed to ignite the tank would be present. "If there is separation of the wires there's no problem. But bundled wires should be a no-no," was his assessment of the potential hazard.

As the crash investigation continued, the FAA agreed. In 1998, the agency ordered airlines to modify the tank wiring in U.S.-registered 747s so that the FQIS wiring would never run with other aircraft wiring. They were given three years to comply. The rule does not apply to all 747s because Boeing started isolating FQIS wiring in 1989, according to Steve Hatch, who was chief project engineer for the model until his retirement in July 1999.

"We really went after separation when we started with autopilots and we had to have redundancies, multiple sys-tems," Hatch recalled. "You want to have separation. You don't want to have other wires that can short to them and cause something in the flight controls."

Still, Boeing initially dismissed the possibility that a short circuit into the FQIS wiring could have caused the explosion on TWA Flight 800 in the summer of 1996.

"If the high and the low come together there would be a circuit breaker," said Rich Breuhaus, Boeing's project engineer for fuel system safety at the time of the crash.

"If the wiring inside the tank has failed in a manner that it shorted, the FQIS will not work properly." And, he added, "Even on an older airplane we have never had a short circuit through the FQIS wiring."

Breuhaus's argument, that such problems would have symptoms, rang a bell with investigators. There had been symptoms. They recalled several inexplicable electrical events in the last few minutes of Flight 800. Individually, they might not have drawn much attention. In the light of what happened, investigators had to question if they signaled the explosion to come.

Before the plane took off, the fuel loader at JFK had to pull a circuit and override a refueling system shutoff because it was not working properly. The volumetric shutoff uses the FQIS system to gauge when the tank is full.

During the brief flight, Captain Ralph Kevorkian had twice mentioned a crazy fuel flow indicator. Wiring to that cockpit indicator is routed with the FQIS wiring, and it runs with 350-volt wires that power cabin lighting.

Finally, when the cockpit voice recorder stopped, it was the microsecond of noise at the end of the tape that got all the press. But what caught the attention of CVR technician Jim Cash were a couple of interruptions in the background signal nearly two seconds earlier, a "discontinuity" in the recording is how Cash described it, and it indicates a brief drain on the electrical power to the CVR.

"It's in a bad spot right near the end. It could be a pump coming on or something going on in the galley, something that put a load on the airplane," Cash speculated. "It's indicative of a high load, something drawing a high amount of current."

Were all these events related? It was time for the crash detectives to follow the advice Dr. Birky had given his group members earlier in the investigation. Don't get lost in the trees, look at the forest, he had told them. Don't look at individual pieces, look at the whole picture.

24

For a few days in the spring of 1997, the microwave oven in the lunchroom at NTSB headquarters was not working. Taped to the door of the appliance was a note that read, "Out of Service." In smaller letters below, some eager investigator carefully noted what had caused the appliance to go kaput. It is an example of the inherent inquisitiveness that afflicts many of the investigators at the safety board. The NTSB has looked into 100,000 airplane accidents since it was created in 1967. Most of them involve small aircraft; most of them, while not uncomplicated, are at least explicable. TWA Flight 800 seemed likely to end up as one of the few major airline disasters that would not be solved, despite the involvement of so many people with an inexhaustible desire to know what happened.

In the fall, investigators made the most important discovery in the search for the source of the ignition of TWA Flight 800's fuel tank. A small black mark was found on a fuel probe connector in the wreckage of the plane. It was identified as a bit of copper-sulfur, created over time by the interaction of the metal conductor with the sulfur in jet fuel. It's the same chemical process that makes a silver spoon tarnish when it's dipped into sulpherous egg yolk.

As he called around trying to determine the significance of the residue, the NTSB's Bob Swaim was told to call George Slenski, a civilian scientist at Wright Patterson who'd had an interesting experience with this material seven years earlier. Some fuel tank wiring with a similar buildup had been taken off an Air Force plane and brought to the maintenance bench for testing. As the worker put an electrical tester to the wiring, the sulfides quickly burned off, igniting the fuel vapors in the cavity of the component with a dramatic *poof!* Such a test performed in the fuel tank would have blown the technician and the plane to bits. Slenski was called to have a look.

He learned from this experience that the residue was semi-conductive though extremely fragile. It could sit harmlessly on fuel tank wiring for years. Maybe forever. But subjected to a sudden jolt of electricity, it would burn off, fast and hot. How much voltage would it take before sparking? How much energy was given off in the spark? He didn't know. Information on the conductivity of sulfides was scarce.

Swaim shared with Slenski that Boeing technicians were also investigating the conductivity of copper-sulfides. Their early test on probe wiring removed from a 747 getting a new FQIS system showed that the deposits could not carry a charge. In the course of removing the probes, the residue had been disturbed. Swaim wondered if that had affected the outcome. They decided new tests should be conducted, this time with copper-sulfur residues intact. But where could they get such wiring?

More than two years after the crash, Swaim got a telephone call from the FAA. Conscientious maintenance workers for Tower Air, an all-747 carrier based in New York, had indications of a short circuit in a center fuel tank on one of their planes.

When the workers removed an FQIS probe, they found the same black deposits on a small part called a terminal strip, where the probe wiring terminates. It was the chemical residue so interesting to Flight 800 investigators. Tower Air turned the prize over to the FAA, and officials called Swaim, who would only have been happier had he'd won the lottery.

For months, Swaim kept the precious terminal strip carefully swaddled in a plastic case so as not to damage the thin film of sulfides. He and Birky went so far as to consult a chemistry lab to get a more exact reading of what the material was composed of. Perhaps they could produce the chemical themselves and perform as many tests as they pleased.

Unfortunately, the residue was a complicated mix of organic and inorganic materials. How it developed on the wiring, its density, texture, and age seemed to play a role in how electrically conductive it was. They left the lab feeling that they'd better take good care of the sample they had. Swaim had every intention of doing so. Unlike the probe wiring Boeing had examined, he knew the residues on this piece from Tower Air had been causing real problems in the center tank of an active passenger-carrying 747. He wanted to gather as much advice as possible before doing any destructive testing. "I didn't want to make a mistake with it," Swaim explained.

When Slenski and another Wright Patterson scientist named David Johnson finally got their hands on the terminal strip in the spring of 1999, there was a lengthy list of exams the safety board wanted them to perform on the small sample. Since the tests would be destructive, not all of them would be done.

Swaim got an answer that told him the investigation was on the right track when 170 volts applied to the wiring caused the residue to flash and make a popping sound that could be heard clear across the lab. There was no measurement taken of the energy emitted that day, but the scientists were certain it was more than enough to ignite the vapors aboard TWA Flight 800.

For the first time since the baffling investigation had begun, a real pathway for an ignition source in the center tank had been found.

· · ·

On July 15, two days before the first anniversary of the crash, the safety board conducted a series of test flights. The heavily instrumented 747-100 was not originally scheduled to be flown at such a significant time, but planning turned out to be extremely complicated. By the time investigators were ready, it was May. By waiting until July, they were able to fly in weather conditions very much like those on the day of the crash.

The duplication of Flight 800 took off within one minute of the actual flight and took just ten seconds longer to reach the altitude at which the explosion occurred. At that point, the temperature in the center tank was 120 degrees, the temperature in the vapor space, 127 degrees. Investigators assumed conditions were similar on Flight 800. By now they had figured out why.

The large center tank is perched above a bay housing three air-conditioning packs. Pan Am's John Borger recalls the location was selected thirty years before because "it was a convenient place to put them and it maximized cargo space." According to Borger and Boeing's Ivor Thomas, there was little concern at the time about what the heat packs would do to temperatures in the tank.

"I don't think anybody in the Boeing company was concerned about temperatures in the pack bay from a safety point of view," Thomas said of their thinking in the late sixties. Yet less than a decade later, the Air Force discovered high center fuel tank temperatures on the military 747s was causing a critical safety problem. It went right to Boeing with the news.

Air Force 747s, known as E-4s, are intended to serve as airborne operations and communications centers in times of national emergency. One E-4 is always on alert somewhere, revved and ready to fly. Shortly after the E-4s went into service in 1979, the Air Force complained to Boeing that as the plane sat with engines and air-conditioning packs operating, fuel in the center tank was getting so hot it was vaporizing, threatening engine shutdowns. The Air Force was worried and asked Boeing to investigate the problem and come up with solutions. By 1980, Boeing had prepared a four-volume report suggesting procedural changes and ground-based cooling techniques.

After the crash of Flight 800, the safety board told Boeing to provide it with any information it had on fuel tank heating problems. The E-4 study was never mentioned. Nor was it brought to the attention of investigators six years earlier when they were perplexed about what caused a fuel tank explosion that killed eight people on a Philippine Air Lines 737. No, the NTSB learned of the E-4 problem in the spring of 1999 by sheer happenstance, when someone in the air force mentioned it to a safety board employee at a meeting in Oklahoma. Investigators were eager to learn more, but getting the document was difficult.

Investigators still hadn't been able to get their hands on a copy when, on May 17, 1999, an NTSB official called both Boeing engineer Rich Breuhaus and former Boeing fuel systems expert Ivor Thomas, who had recently joined the FAA. Both Breuhaus and Thomas explained that they learned about the E-4 study at the same air force meeting. Neither man said why, even then, Boeing did not pass the information along to the safety board.

Boeing insists the commercial side of the company was never told the military side was conducting a study for the air force. Yet the report itself suggests that there must have been some communication. A similar fuel tank heating problem on a commercial jet is discussed involving a Japan Air Lines 747 that lost fuel flow from the center tank as the jumbo took off from an airport in Hawaii.

By July Boeing officials were summoned to Washington, where an agent with the General Accounting Office threatened that the company could be charged with obstructing a federal investigation, though later the GAO concluded there was no evidence of an intentional cover-up. The issue still raises the blood pressure of some NTSB investigators.

It could be argued that Boeing should have made an effort to learn more about the conditions in the center tank in light of more than three decades of debate over whether the isolation of ignition sources was sufficient to prevent fuel tank explosions.

"It's so difficult to ignore the fact that you can take off and fly with a fuel tank bomb," says Charles Anderson, who worked for Boeing developing fuel tank inerting technology. In fact, it wasn't difficult to ignore at all. Until Flight 800, it was a well-kept secret that industry and the FAA had tallied up the fuel tank explosions in the past, divided by the number of total flights, and deemed the hazard an acceptable risk.

"People are suddenly saying to themselves, 'My goodness, you mean we can carry fuel around that can explode?' Well, we discovered that forty years ago," said the FAA's Tom McSweeny, explaining that, "At the time we did not see a critical safety issue."

In the case of Flight 800, the significant discovery was how often the heat sources below the 747 center fuel tank would send it into the explosive range. The issue then became how to deal with it.

An important first question for the board's explosion expert, Merritt Birky, was whether the heat from the air packs was radiant—that is, did it come off the packs in waves—or convective—did it move with the air flow.

"You want to keep the tank cool. If it's convective, ventilation will help keep it cool. Radiated heat can be reflected back." Yet Birky felt Boeing was resisting trying either method of cooling the packs. In the spring of 1998, he realized why.

Birky, Loeb, and other safety board staffers were meeting with Boeing's Thomas, 747 chief project engineer Steve Hatch, and fuel systems engineer Kevin Longwell, kicking around different suggestions for reducing the temperature in the fuel tank. Once again a suggestion was made to lay insulation between the air packs and the tank.

Only this time, Thomas explained that the heat from the pack bay was transferred to the tank by design. We use the tank as a heat sink, he explained.

Loeb remembers he was "dumbfounded" by the remark. Birky also felt as if he'd been duped. They were two years into the investigation, and Boeing had never before told them that redirecting the heat away from the tank couldn't be done without damaging the air-conditioning packs. Looks were exchanged among a number of the safety board staffers attending the meeting, one asking, Are you saying that you can't insulate because you need to have the tank absorbing the heat from the pack bays?

"To us it was an obvious thing to see," Thomas said later. "I didn't understand why he was getting emotional about it" he said, referring to Loeb. "It was like, what's the big deal?"

Loeb did think it was a big deal, and he wasn't alone. Among themselves, the safety board staffers expressed their amazement at the news and their surprise at the design, Loeb offering his opinion that "designing an airplane with a fuel tank that is a heat sink made no sense whatsoever."

This design is not an uncommon practice, though, according to Richard Hill, the FAA's long-time aircraft fire safety program manager.

"Normally, there are a number of airplanes where the heat source is the air-conditioning and the tank above is a heat sink."

Fred Grenich, a twenty-two-year veteran of fuel systems at Boeing at the time of the crash, explains why. "Fuel is a BTU garbage dump. Dump the heat into the fuel and it's not a problem." The problem arises when there is little or no fuel in the tank.

After the disaster, while Grenich worked on the special in-house investigation into the circumstances surrounding Flight 800, his opinion changed.

"The air-conditioning pack under the fuel tank isn't a great idea," he concluded. Others on the task force were agreeing with him. Even Ivor Thomas. "When we looked critically at the air-conditioning packs under the fuel, it became evident it was a potential problem."

It was a problem specific to Boeing airplanes. All seven models of Boeing airliners share the design that places heat-generating air packs beneath a center fuel tank. The McDonnell Douglas airplanes that became part of Boeing's product line when the two companies merged in 1997 do not. Airbus Industries also has air-conditioning packs under center fuel tanks, but a ventilation system successfully minimizes heating of the tank.

"Fuel temperatures were on our minds all the time because the engineers designing Airbus planes had been in on the design of the Concordes," said Marten Bosman, director for flight safety for Airbus Industrie. Supersonic flight generates heat from friction. Airbus had to develop methods of keeping temperatures down on the Concorde's fuel tanks.

Pan Am and TWA's voluntary switch from Jet B to Jet A following the 1963 Pan Am crash in Elkton, Maryland, was supposed to provide extra protection against fuel tank explosions. What Flight 800 suggests to experts like Bernard Wright, a fuel specialist with the Southwest Research Institute in San Antonio, is that the safety benefit of Jet A over Jet B is lost if vapors in the tank regularly rise above the flammability limit.

"It doesn't make any difference what type of fuel it is because when it's above its flash point there's no difference in the ignition energy."

It was a point well taken by the FAA. Another committee, the Fuel Tank Harmonization Working Group, was formed to figure out ways to reduce the time a fuel tank is in the flammable range. The report, released in the summer of 1998, suggested that center tanks with adjacent heat sources were explosive about thirty percent of operating time, four times more often than wing tanks.

An appropriate goal, according to the representatives on the committee, would be to "reduce the flammability levels in center tanks" to the amount of time that wing tanks were flammable, about seven percent of the time.

The dilemma, predictably, was how to do it at a price industry was willing to pay. In the cost benefit analysis, the cost of doing nothing, in other words, the cost of any future accidents, was figured as about $2 billion over the next ten years. Every solution examined cost more.

Critics and insiders alike raised their eyebrows over the numbers, one safety expert offering that the report "basically uses numbers to strengthen the opinion that there's nothing that can be done."

Boeing's Rich Breuhaus insists the company is working to see what can be done to make things better, though at the same time he is ticking off as unworkable a list of suggested fixes.

Circulating cooler air in the narrow space between the packs and the tank as Airbus does had only a "modest effect" on tank temperature, according to Breuhaus.

The problem with switching to a fuel with an even higher flammability level than Jet A was a lack of availability. There is also a concern that it could make engines harder to start at cold temperatures.

As for inerting, Breuhaus characterized the military's experience with OBIGGS as "very onerous." The cost could not have escaped his attention, either. The harmonization group estimated the cost of OBIGGS at $34 billion. A far simpler plan for inerting tanks on the ground before takeoff would cost between $3 billion and $4 billion.

"Boeing will use Band-Aids and ignore the problem as long as it can because billions of dollars are at stake here," says aviation lawyer Mitch Baumeister. "If it can contain the catastrophes, it may be able to prevent the financial lid from blowing off this problem."

The safety board doesn't have to consider cost in making recommendations, and it clearly did not when it made its request that commercial airliners be prohibited from flying with explosive fuel-air mixtures in the tank. It took more than three years, but in the fall of 1999, the FAA followed up on the safety board's recommendation and ordered a review of fuel tank design on commercial aircraft. FAA Administrator Jane Garvey was calling for a "fundamental change in how fuel tanks are designed, maintained, and operated." How this was to be accomplished was not clear, but modifying fuel tank designs presents Boeing a daunting task. Every model airplane the company produced in the jet age, from the 707 to the 777, could be affected, a total of 8,000 airplanes in 1999. The cost of solving such a broad-based problem is staggering. With so much at stake, Boeing will try to influence the process every step of the way. Some charge it already has.

Shortly after the NTSB's public hearing on the crash of Flight 800, when it became clear that the FAA would have to do something to address the hazard of fuel tank explosions, the agency's director of certification and regulation, Tom McSweeny, stunned many in the industry by hiring Ivor Thomas to be what he called the FAA's "guru in fuel tanks." Thomas had disregarded the need for fuel system explosion protection mechanisms from the time he arrived on Boeing property in 1966. Now he would be the person air safety regulators turned to for guidance.

25

For his role in the theft of TWA crash wreckage, Captain Terry Stacey pleaded guilty to misdemeanor theft charges in Federal court. A letter he had received from Pope John Paul II, whom he'd flown on a TWA 747 in September of 1987, was presented to the sentencing judge, who gave Stacey three years probation and a fine. He returned to TWA as a pilot and testified against James and Elizabeth Sanders, who were tried and convicted in April 1999 and sentenced to probation and community service on the eve of the third anniversary of the crash.

Ian Goddard posted a note on the Internet in November 1997 calling his earlier efforts to pin the crash on the Navy "reckless and a mistake." Pierre Salinger, saying he'd taken a beating in the press because of his pursuit of the missile theory, was no longer going to pursue the issue, either.

It would be wrong to dismiss the missile theorists as simply out of touch or paranoid; there were so many inconsistencies in the official statements. Many of the contradictions could be attributed to the unusual circumstances of the crash, the chaotic response, and the conflicting agendas of the major agencies involved. Years after the crash, however, even with the benefit of hindsight, the most active proponents of the theory that the U.S. government covered up the true cause of the crash still do not accept that bungling, benign or otherwise, explains all the misstatements.

Captain Russell surrendered his radar videotape to FBI agents rather than face a grand jury. He has not surrendered his commitment to expose what he still holds to be a government cover-up.

"They've stifled this entire investigation," he says. "I'm at loss to know what to do next. But when I see something that I can do, I'm willing to do it."

If there was conspiracy to hide the truth about what really happened to Flight 800, it began long before the crash, and it wasn't an effort to cover up action, it was a conspiracy of inaction. Although the newspapers were filled with experts insisting planes just don't fall out of the sky and fuel tank explosions can't bring down an airplane, insiders knew different.

On a rainy December in 1963, Captain Eugene Banning stood over the wreckage of Pan Am 214 and made a silent promise to his fellow pilots and seventy-seven others. He was still trying to keep that promise when he ran into NTSB chairman Jim Hall at an air safety conference in Barcelona in 1998. The two men discussed the TWA 800 investigation and Banning's work on all those fire safety committees decades earlier. "I said he should find the minutes of those meetings and read up on it. It would help his understanding of the history of this thing," Banning said.

No one in commercial aviation likes to talk about it publicly, but safety improvements are nearly always viewed as a balance. How much will they help? How much will they cost? When the cost outweighs the benefit, certain hazards are considered an acceptable risk.

David Houck, fifty-one, dismissed the missile theory early on and paid close attention to the various ignition scenarios being pursued by the NTSB and the solutions recommended. His sister, Susan Hill, died in the crash of TWA Flight 800.

"I told my mom, 'The only thing I know is what Susan would do if it was me that went down. She'd go after them. She'd want to do everything that could be done.'" That's because Hill was a champion for the underdog. Houck knew that about his kid sister from growing up with her and working with her at the Portland Police Bureau in Oregon.

Susan was a homicide detective. Houck was an bomb technician. He understood the safety board's investigation and how a small spark could bring down the giant jetliner. What he does not understand is how a financial calculation can be used to address the safety problem reexposed by the TWA disaster.

"For them it's the bottom line," he said, "but I wonder what those number crunchers would think if they lost their wife and kids because someone decided it would be more expensive to fix something than to ignore it?"

TWA Flight 800 was the fourteenth fuel tank explosion on a commercial airliner in thirty-five years. Each and every time, industry and regulators dealt with the ignition source and ignored the explosive tank. Had the cause of the ignition of Flight 800 not been so elusive, the disaster might have been handled the same way.

"No one wants to know the cause of the crash more than I do," insists retired Boeing engineer Steve Hatch. He was there in 1969 when the first 747 was rolled out. The last three years of his career were absorbed by the most troubling event in the 747's history.

"There's one pump we didn't find, and God, I wanted that pump so bad. I almost got to the point I wanted the pump to show the pump was bad so we could fix it and get on with it."

Fix it and get on with it. That's how fuel tank explosions have been handled. "There's super emphasis on finding the cause of the accident, says C. O. Miller, an aviation safety expert and former NTSB investigator. "But there are many factors."

One consequence of Flight 800's mysterious fall into the sea was challenging the symptomatic approach to fuel tank explosions.

"It would be a wonderful feeling if we could identify it," said the NTSB's Bernie Loeb, referring to the source of the ignition, "and make a recommendation to eliminate whatever it is. But it wouldn't eliminate the potential for further explosions." On that he is emphatic.

John F. Kennedy Jr.'s small plane flew into the Atlantic near Martha's Vineyard on July 16, 1999, killing him; his wife, Carolyn Bessette Kennedy; and her sister, Lauren Bessette. As the families of those who died on TWA Flight 800 gathered under a striped tent at Smith Point Park on Long Island the next morning for a memorial service for their loved ones, they noted the odd coincidence that the event that had consumed their lives had collided with a new tragedy consuming the nation.

There were about 100 people at the ceremony, fewer than in previous years. There was crying, but not as much. There were curious onlookers, but not as many. Even the media, which had been so voracious three years before, had disappeared to cover the Kennedy story. In the next day's papers, the anniversary would barely get a mention.

Hall spoke at the service, but otherwise he was on the phone getting the Kennedy crash investigation going, including dispatching to Martha's Vineyard some of the same people who had worked on Flight 800. Afterward, Hall and the families went to the hangar where the reassembled 747 was still on display. Ron, Margret, and Chris Krick stood by the cockpit area, where Oliver's seat and the flight engineer's desk were set up.

Always the pilot, Ron found himself talking to Hall about the Kennedy crash.

A plane falls from the sky, and there are always questions. How could it have happened? What could have gone wrong? The Kricks and Hall were standing in the shadow of a monument to the effort made to answer those questions.

Hundreds of scenarios were explored, thousands of experts consulted, and millions of pieces of information analyzed. All to an end that seems to say the plane could have exploded because of a microscopically small buildup of chemicals on tiny wires in the heart of the plane. It was amazing that such a minute problem could be discovered at all. It was as close to a "Eureka" moment as Swaim, Birky, and the other Flight 800 investigators might ever get.

But it exposed the fallacy of four decades of assurances that ignition sources can be kept out of fuel tanks. If there is any legacy to TWA Flight 800, it is that such an approach is hubris and, more times than forgivable, dead wrong.

BIBLIOGRAPHY

Banning, E. R., "Problems and Procedures of Central Safety Chairman," presentation to ALPA Air Safety Forum. Chicago, October 6–8, 1964.

Barlay, S. *The Search for Air Safety.* New York: William Morrow, 1970.

Barlay, S. *The Final Call.* New York: Pantheon Books, 1990.

Brookley, W. *Fuel Tank Nitrogen Inerting Tests.* Dayton, OH: USAF, 1970.

Brookley, W. *Historical Information on Fuel Tank Explosion Protection for Aircraft,* written for Society of Automotive Engineers, not published, 1998.

CAB Aircraft Accident Report, Pan American World Airways Flight 214. Washington, D.C.: U.S. government, 1965.

Cahill, P. *Electrical Short Circuit and Current Overload Tests on Aircraft Wiring.* Atlantic City, NJ: FAA Technical Center, 1994.

Coordinating Research Council. *Aviation Fuel Safety.* New York: 1965.

Department of Transportation. *FAA Report to Congress Systems and Techniques for Reducing the Incidence of Post-Crash Fuel System Fires and Explosions.* Washington, D.C.: Department of Transportation, 1988.

Eddy, P., E. Potter, and B. Page. *Destination Disaster*. London: Hart-Davis MacGibbon, 1976.

Emerson, S. "Pan Am Scam." *Washington Journalism Review,* pp. 15–20, 1992.

Federal Aviation Administration. *Fuel Tank Harmonization Working Group Final Report*. Washington, D.C.: FAA, 1998.

Fielder, J. and D. Birsch. *The DC-10 Case*. Albany: State University of New York Press, 1992.

Fox, F., M. S. Sisk, and R. DiBartolo. *Immersion Studies of Aircraft Parts Exposed to Plastic Explosives*. Atlantic City, NJ: Federal Aviation Administration, 1996.

Fredrick, S. *Unheeded Warning*. New York: McGraw-Hill, 1996.

Fuel Flammability Task Group. *FAA Review of the Flammability Hazard of Jet A Fuel Vapor in Civil Transport Aircraft Fuel Tanks*. Washington, D.C.: Federal Aviation Administration, 1998.

Gandt, R. *Skygods*. New York: William Morrow and Company, 1995.

Haack, G. "Fuel System Inerting With On Board Inert Gas Generating System (OBIGGS)." Presentation to SAE and FAA, Washington, D.C., October 7–9, 1997.

Haddad, J., W. McAdoo, and O. Ball. *Service Experience With Liquid Nitrogen Fuel Tank Inerting System in FAA DC-9 Aircraft*. Washington, D.C.: Federal Aviation Administration. 1972.

Heppenheimer, T. A., *Turbulent Skies*. New York: John Wiley & Sons, 1995.

Hill, R. and G. Johnson. *Investigation of Aircraft Fuel Tank Explosions and Nitrogen Inerting Requirements During Ground Fires: Final Report*. Atlantic City, N.J.: Federal Aviation Administration, 1975.

Ingells, D. *747: Story of the Boeing Super Jet* Fallbrook, CA: Aero Publishers, Inc.: 1970.

Irving, C. *Wide-Body: The Triumph of the 747*. New York: William Morrow and Company, 1993.

Johnston, M. *The Last Nine Minutes*. New York: William Morrow and Company, 1976

Kuter, L. *The Great Gamble: The Boeing 747*. Tuscaloosa: The University of Alabama Press, 1973.

Lessons Learned Newsletter. Directorate of Aerospace Safety, Volume 5, Item 1, Norton AFB: California, March 1977.

National Transportation SB Docket No. SA-516. Washington, D.C., 1997.

Newhouse, J. *The Sporty Game*. New York: Alfred A. Knopf, 1982.

Norris, G. and M. Wagner. *Boeing 747*. Osceola, WI: Motorbooks International, 1997.

North Atlantic Treaty Organization. *AGARD Lecture Series No. 123*. Neuilly-sur-Seine, France, 1982.

North Atlantic Treaty Organization. *AGARD Advisory Report No. 123*. Neuilly-sur-Seine, France, 1979.

North Atlantic Treaty Organization, *AGARD Advisory Report No. 132*. Neuilly-sur-Seine, France, 1979.

O'Flaherty, J. *Handling Catastrophe Despite Official Help*. Airport Press, 1996.

Preston, E. *Troubled Passage*. Washington, D.C.: U.S. Department of Transportation, 1987.

Rogers, E. *Flying High*. New York: The Atlantic Monthly Press, 1996.

Special Aviation Fire and Explosion Reduction (SAFER) Advisory Committee. Washington, D.C.: Federal Aviation Administration, 1980.

Sanders, J. *The Downing of TWA Flight 800*. New York: Zebra Books, 1997.

Serling, R. *Howard Hughes' Airline: An Informal History of TWA*. New York: St. Martin's Marek, 1983.

Serling, R. *The Jet Age*. New York: Time-Life Books, 1982.

Serling, R. *Legend and Legacy: The Story of Boeing and Its People*. New York: St. Martin's Press, 1992.

Sumwalt, R. "Avoiding the Fate of Icarus," *Air Line Pilot Magazine*, April (1997): pp. 10–13.

Trans World Airlines, Annual Report. St. Louis, MO: 1995, 1996, 1997.

Wright, B. and D. Zallen. *Assessment of Concepts and Research for Commercial-Aviation Fire-Safe Fuel.* San Antonio, TX: Southwest Research Institute, 1998.

INDEX

Abels, Mark, 73–74, 83, 85, 94–95
Abyss, The (Card), 18
Acceptable risk, 176, 228, 231–32, 234–35
Achilli, Craig, 41, 44
Adak (Coast Guard cutter), 41, 43
Aeromedical Evacuation Squadron, McGuire AFB, 104
Air Disaster Alliance, 218
Air handling units, 13, 27, 201–2, 226–30
Air Line Pilots Association, 177
Air Transport Association, 88
Airbus Industries, 230, 231
Airline disaster response, 86–89
Alexander, Matthew, 17, 27, 29
Alexy, Paul, 66–67
Alitalia Flight 611, 82
Allegheny Airlines, New Haven crash, 1971, 177
Allen, Bill, 122
Aloha Airlines 737, 1988 structural failure, 153, 216, 217

American Eagle Flight 4184, 1994 crash, 147–48
Anderson, Charles, 179–80
ATF (Bureau of Alcohol, Tobacco, Firearms), 61, 205–6
Avianca Airlines 727, bomb and fuel tank explosion, 162, 204
Aviation Consumer Action Project, 177
Aviation Disaster Family Assistance Act, 88

Banning, Eugene, 68–69, 123–24, 175, 177, 234
Barham, Elena, 9, 169, 188
Barlay, Steven, 65
Barr, Michael, 193
Baumeister, Mitch, 231
Baur, Christian, 37–39, 103, 132–33
Bellazoug, Jasmine, 11
Bellazoug, Myriam, 11
Benzon, Bob, 162

Bessette, Lauren, 236
Birky, Merritt, 62–63, 155–56, 158,
    163–65, 172–73, 186, 187, 190,
    197, 202, 219, 225, 228, 229, 237
Black boxes, 59–60, 111–12
    analysis of, 118–19
    cockpit voice recorder (CVR), ix,
        26, 92–93, 154, 220–21
    flight data recorders (FDR), ix,
        92–93
    Forbes erroneous
        announcement of finding,
        116–17
    NTSB excellence at retrieval, 92
    recovery of, 117–18
Block, Ed, 56–57, 216–18, 219
Boak, Naomi, 12
Bodies, passengers and flight crew
    autopsy of, 34, 166–67
    location and recovery of, 39, 43,
        45, 46, 51, 52–53, 75,
        112–14, 118, 157, 171
    morgue, Hauppauge, New York,
        51, 93, 107
    personal effects and
        identification of, 50, 93,
        106–7, 114–15, 157, 165–67,
        168, 169
    pilots, 31, 75
    underwater, in plane, 60, 114,
        115–16, 157
Boeing Corporation
    acceptable risk and costs, 176,
        228, 231–32
    certification of 747 fuel system,
        190–91
    development of 747, 121–23,
        215
    development of SST, 215
    E-4 study of fuel tank heating
        problems, 226–27

FAA-ordered changes in 747
    fuel system, 191–92
factory-issued service bulletins,
    191–92
fuel system inerting, rejection
    of, 124–29, 175–76, 178, 231
ignition sources and fuel tank
    design, 190, 223–25, 236,
    237
media campaign on flight
    safety, 201
oldest commercial designs of, 2
and Pan Am, 122–23, 215
test of fuel tank explosion,
    201–2
Boeing 707, 121
    crashes of, 65–69, 71, 126
    fuel system, 124, 126, 129
Boeing 737, 2, 87, 155, 156, 181,
    216
Boeing 747
    assembly line, Everett,
        Washington, 2, 122, 123, 127
    development of, 121–23,
        127–28
    E-4 (military 747) study of fuel
        tank heating problems,
        226–27
    FAA-ordered changes in wiring,
        219
    fuel system design, 123–29,
        156, 190–94
    fuel tank explosion test, 201–2
    image of, 7, 15
    inspections of, 2
    Japan Air Lines 747, center tank
        problem, 227
    life expectancy, 215–16
    third crew member required, 3
    Tower Air, all-747 carrier,
        224–25

TWA fleet, x, 19–20, 216
wiring, 2, 57, 215–19
Boeing 757, 19, 20
Boeing 767, 19, 20
Boergesson, Bob, 48
Borger, John, 121, 226
Borjesson, Christina, 212–14
Bosman, Marten, 230
Boyd, Alan, 69
Brady, Shaun, 37, 44–46, 133, 134
Breuhaus, Rich, 220, 227, 231
Brookley, William, 129, 175–76,
177, 178
Busick, Paul, 138–40
Buttaroni, Katia, 18
Buttaroni, Mirco, 6–7, 18, 29
Buttaroni, Monica Omiccioli, 6–7,
18–19, 29, 156–57, 165

C-5 military transports, 178
Campbell, Dan, 205–6
Campbell, Richard (flight
engineer), 2, 3, 21, 30–31
Cash, Jim, 119
Challenger explosion, 62
Charles de Gaulle Airport, Paris,
49–50
Christopher, Charles, 40, 53
Christopher, Janet, 40, 53
Civil Aeronautics Board (CAB), 69,
69n, 70–71
Clapp, Chris, 35–36
Clinton, Hillary Rodham, 171
Clinton, William, 85, 99, 170–72,
218
Aviation Disaster Family
Assistance Act, 88
Clodfelter, Robert, 179–80
CNN, x, 36, 108, 141–42, 145,
154

Coast Guard Station, East
Moriches, 40–41, 48–49, 50, 60,
93
imposter at, 101–4
makeshift morgue at, 50–51
media at, 108–9
search and rescue, Flight 800
crash, 35–36, 40–42, 43–44, 97
Concorde, 127, 230
Cone, Jim, 79
Conspiracy theory. *See* Missile
theory
Corrigan, Thomas, 102
Cover-up allegations, 116–17,
233–34
Crash investigation, 43, 97–99,
149–51. *See also* FBI, NTSB
ATF, 61, 205–6
FBI, 53–55, 57, 60, 97–99, 133,
137, 143, 150, 154–55
FBI-NTSB rivalry, 56, 98–99,
148–51, 161–63, 172–73,
205–6
ignition source found, 223–25,
236
inspections of inside of fuel
tank, 156, 157–59, 189–90
NTSB, 59–61, 62–63, 97–99,
117–18, 137, 139–40,
155–56, 189–94, 223–32
rebuilding of aircraft at
Calverton, Long Island,
197–200, 213
red foam identification, 210–11
remains of aircraft and reassembly at Calverton, Long
Island, 148–49, 156–58,
162–63, 205
residue of chemical explosives
found, 163–65, 172–73,
185–86

Crash investigation, *continued*
   suspicions and secrecy, 210–11
   test flights, 747–100,
      duplication of Flight 800,
      226–27
Crash site, 42, 45, 52–53, 60, 75
   bodies, 39, 43, 45, 46, 52–53,
      60, 75, 118
   debris, 29, 38–39, 42, 43–44, 57,
      97, 186
   depth of, 60
   fire at, 42, 43–44
   investigation at, 111–12, 117
   location of 747 tail, 112
Craycraft, Kenneth, 216
Cremades, Ana (Vila), 18, 183–84
Cremades, Daniel, 17–18, 27, 29,
   183–84, 200
Cremades, Dario, 18
Cremades, David, 18
Cremades, Jabina, 18
Cremades, Jose, 18, 183–84, 200

Daniels, Denise, 56
Danielson Holding Corporation,
   10, 188–89
Darcy, Kevin, 210
Darden, Joseph, 84
DC-8, Anchorage crash, 1970, 177
DC-9, 22
   inerting system installed on,
      176, 177
Deadheading, 3
Deane, Kelvin, 186–87
Defense Industrial Supply Center,
   Philadelphia, 216
Dickey, Deborah and Douglas, 6,
   29
Dickinson, Alfred, 56, 59–60, 99,
   150, 153, 162, 186

Directorate of Aerospace Safety,
   adoption of nitrogen inerting
   systems for military, 178
Dole, Elizabeth, 180
Donaldson, Lufkin and Jenrette, 10
Donner, Bud, 137
Dorfman, Gerald, 203–4
Dunsky, Ron, 141–43
E-4 study of fuel tank heating
   problems, 226–27
East Moriches Fire Station, 101–2.
   See also Coast Guard
Eastwinds Airlines Flight 507,
   32–33
Eckrote, Debra, 192
EgyptAir 767 crash, 206–7
Erickson, Jeffrey, 73–74, 83, 85,
   94–95, 169

F-14 (Navy Tomcat)
   accidents, 57
   wiring, 215
F-104 test explosion, 125
FAA (Federal Aviation
   Administration)
   aging structure inspection
      program, 2, 214
   Aging Systems Task Force, 218
   committee on fuel tank
      inerting, 124
   fears of insufficient security,
      139
   and Flight 800 investigation,
      163
   Fuel Tank Harmonization
      Working Group, 230–31
   inerting of fuel tanks, testing
      of, 176–77
   and missile theory, 63–64,
      137–40, 163

NTSB indictment of for
American Eagle Flight 4184
crash, 147–48
-ordered changes in 747 fuel
system, 191–92
-ordered changes in wiring, 219
proposals on inerting, 177–79
recommendations on inerting,
70–71, 180, 204–5
requirements on inerting,
177–78
review of fuel tank design, 232
rules, cockpit conversation, 224
standby plane, 55
Technical Center, 63–64
tombstone regulation, 71
ValuJet crash as public relations
disaster for, 171
Families of Flight 800 victims
area at Kennedy set aside for, 77
Clinton meets with, 170–72
criticism of Suffolk County
medical examiner, 166, 170
criticism of TWA, 113–14, 170
and delay in TWA response, 82,
89
grief of, 79, 96
Giuliani reads passenger list to,
82–83
Giuliani role as family
advocate, 93–94
housed at Ramada Plaza Hotel,
78, 81, 93, 170–72
identification and claiming of
loved ones, 165–67, 169–70
Paris, 95–96
public support of, 151–52
questionnaires for, 167–68
social services available to, 94
and TWA's Johanna O'Flaherty,
88–89

vicarious traumatization, 151
visit to plane reconstruction at
Calverton, Long Island,
198–200
Wetli disagreement over access
to, 93–96
Fassett, Adrian, 104
FBI
EgyptAir 767 crash, 206–7
explosive residue found by,
164–65, 172–73, 185–86
help provided to NTSB, 91–92
interagency investigations, 97,
205–7
and Kallstrom, Jim, 40, 53–55,
57, 91–92, 97–99, 114, 137,
143–44, 150–51, 153,
163–65, 172, 185, 197–98,
200–201, 205–6, 209, 213–14
lab, criticism of, 165
notification of, about crash, 40
personnel and equipment on
scene, 91–92
questioning of witnesses,
133–34
rivalry with NTSB, 55–56,
98–99, 148–51, 161–63,
172–73, 205–6
theories about crash, 53–55, 56,
57, 61, 137, 149, 161–62, 163
Feeley, David, 94
Fenwal flame-suppressing system,
126
Finkle, Jim, 133
Flight attendants
analysis of passenger behavior,
ix–x
Christopher, Janet, 40, 53
Hull, Jim, 47
Johnsen, Arlene, 11
Kempen, Michael, 47–48

Flight attendants, *continued*
  Lang, Ray, 19, 20, 23, 47
  Miller, Laura Beth, 47–48
  routine of, 23
  Simmons, Olivia, 47
  Torche, Melinda, 19, 23
Flynn, Cathel "Irish," 139, 185–86
Forbes, Michael, 116–17, 134,
  212–14
Francis, Robert, 55–56, 57, 60, 91,
  97–99, 112, 115, 116, 150–51,
  153, 163, 171, 199
Francis S. Gabreski Airport, 37
Freeh, Louis, 53, 172, 207
Friendly fire or U.S. Navy shoot-
  down theory, 54–55, 109,
  141–45, 209–14, 233
Fuel
  Airbus and SST concerns, 230
  fire-suppressing additive,
    180–81
  flash point and fuel type, 230
  Jet A, 3, 25, 70
  Jet B, 70, 71, 230
  sulfur in, 223–24
  temperatures, 228
  and TWA Flight 800, 3, 24–25
Fuel system design, x–xi. *See also*
  Inerting systems
  adoption of nitrogen inerting
    systems for military, 178
  center tank, 3, 27, 28, 63, 70,
    124, 126, 156, 184, 186–87,
    193, 224, 225
  center tank as heat sink,
    228–30
  cross-feeding, 23
  E-4 study of fuel tank heating
    problems, 226–27
  FAA-ordered changes in 747
    fuel system, 191–92

  factory service bulletins,
    191–92
  flammability, xi
  ignition sources, 190–94, 200,
    204–5, 217–21, 223–25, 236,
    237
  inerting, 70–71, 124–29, 175–81
  pumps, 23, 190–94, 218–21
    235
  ullage, 124
  vapor, 27, 124
  wing tanks, 3, 23, 70, 124–25
Fuel tank explosions, xi, 155,
  234–36
  Allegheny Airlines, New Haven
    crash, 1971, 177
  Avianca Airlines, 727, 204
  awareness of cause and
    prevention of, 175–81
  Boeing test of 747 fuel tank
    explosion, 201–2
  DC-8, Anchorage crash, 1970,
    177
  ignition sources, 192–94,
    217–21, 223–25, 236, 237
  number of, 235
  Pan Am Flight 214 crash, 1963,
    65–69, 70, 124, 125, 177,
    178, 230, 234
  Pan Am–KLM, Tenerife
    collision, 1977, 177–78
  Phillipine Air Lines 737,
    Manila, 155, 227
  TWA Flight 800, crash of 1964,
    71
  TWA Flight 800 theory, 63,
    155–56
  TWA Lockheed Constellation
    crash, Milan, Italy, 1959,
    69–70
Fuschetti, Vincent, 32–33

Gaffney, Robert, 169
Garvey, Jane, 232
Gates, Tiffany, 9
Gimlett, Brian, 143
Giuliani, Rudolph, 75, 82–83,
    93–94, 168
Goddard, Ian, 144, 233
Goelz, Peter, 42–43, 55, 87, 171,
    200, 207
Golden, Bob, 48–49, 50–51, 106, 167
Golden, Kathy, 48
Gore, Albert, Jr., 217–18
Gore, Albert, Sr., 42
Grassley, Charles, 206
Gray, Chad, 10
Gray, Charles Henry "Hank," III,
    9–10, 83, 169–70, 187–89
Gray, Hank, IV, 10, 188–89
Gregg, Raymond and Joan, 68–69
Grenich, Fred, 178, 179, 181, 229
Grossi, Dennis, 118–19
Grumman Aircraft, 148–49

Hahn, Rich, 161, 162, 164
Hall, Jim, 42–43, 55, 86–88, 98, 137,
    143, 151, 172–73, 186, 197–98,
    201, 202, 205–6, 234, 236
Hammerschmidt, John, 207
Harriman, Pamela, 85
Harris, Estee, 87
Hartline, David, 126
Hatch, Steve, 123, 128, 219–20,
    228, 235
Hazle, Shelly, 115
Hendrix, David, 212
Hibberd, Mary, 93, 113, 167
Higgins, Kitty, 98, 116, 139, 171,
    198, 205
Hill, Richard, 180, 229
Hill, Susan, 16–17, 27, 29, 234–35

Hilldrup, Frank, 59, 92, 111,
    117–18
Hinson, David, 139, 163, 194
Hogan, Jamie, 76–77
Horeff, Thomas, 176, 177
Houck, David, 234–35
Huber, Jane Parker, xiii
Huhn, Michael, 157
Hull, Jim, 47
Hurd, Jamie, 199

Inerting systems, 70–71, 124–29,
    175–81
    adoption of nitrogen inerting
        systems for military,
        178
    Boeing rejection of, 126–29,
        175–76, 178, 231
    OBIGGS (onboard inert gas
        generating system), 179–81,
        231
Intini, Frank, 103

Japan Air Lines 747, center tank
    problem, 227
John F. Kennedy International
    Airport, x, 16–17
    Hanger 12, 48, 84, 85
    security, 7
    TWA Constellation Club, 10
    TWA main ticketing lobby, 77
Johnsen, Arlene, 11
Johnson, Craig "Jake," 37, 44, 133,
    134
Johnson, David, 225
Johnson, Jed, 10–11
Jordan, Ken, 42–43
Joshi, Deepak, 59, 61, 92, 111, 186,
    187

Juniper (Coast Guard cutter), 97

Kallstrom, Jim, 40, 53–55, 57,
    91–92, 97–99, 114, 137, 143–44,
    150–51, 153, 163–65, 172, 185,
    197–98, 200–201, 205–6,
    213–14
  terrorist or missile theory of,
        53–54, 57, 62, 163–65, 209
Kallstrom, Susan, 40
Kelly, Mary Anne, 76–77, 79, 83,
    166, 169
Kelly, Michael, 73–75, 76, 78, 83,
    94–95
Kelly, Raymond, 40
Kempen, Michael, 47–48, 105–6
Kennedy, Carolyn Bessette, 236
Kennedy, John F., Jr., 236
Kevorkian, Doug, 22
Kevorkian, Ralph (pilot), ix, 3–4, 8,
    15–16, 20–21, 22, 24, 25, 27,
    30–31, 220
Kimmel, Cleve, 127–28
Knapp, Bob, 102, 103
Knotts, John, 102–3
Knuth, George, 66, 67, 68
Kramek, Robert, 140
Krick, Chris, 22, 236
Krick, Margret, 31, 80, 236
Krick, Oliver "Ollie" (flight
    engineer), ix, 1, 2, 3–4, 8–9,
    21–23, 25–26, 27, 30–31, 79–80
Krick, Ron, 21–22, 79–80, 236
Kristensen, Edward, 143
Kunz, Linda, 211
Kuter, Laurence, 122

L-1011, 21
Laden, Osama bin, 7

La Guardia Airport, TWA,
    Ambassador Club, 76
Lang, Ray, 19, 20, 23, 47
Lawsuits following crash of Flight
    800, 175
Leonard, Joseph T., 193–94
Lesser, Gary, 123
Levine, Naneen, 36, 37
Lightning, as cause of plane
    crashes, 68–70, 126
Loeb, Bernard, 147–48, 154,
    197–98, 200, 201, 206, 228–29,
    236
Loftus, Elizabeth, 134, 135, 137
Long, Eileen, 84–85, 89–90,
    113
Longwell, Kevin, 228
Loo, Patricia, 12, 27, 61
Lothrigal, Attila, 124
Lowell, Vernon, 71
Lowes, Jim, 125, 127
Lychner, Joseph, 84–85, 89–90,
    113–14, 200
Lychner, Kate, 84–85, 89–90,
    200
Lychner, Pam, 84–85, 89–90,
    200
Lychner, Shannon, 84–85, 89–90,
    200

Manno, Joe, 185
Maranto, Stephanie, 199–200
Marlin, George, 77–78, 81–83,
    84–85, 95, 113
Martens, Cynthia, 152
Martz, Roger, 164–65
Mason, Linda, 212
Mawn, Barry, 207
Maxwell, Ken, 153, 161–62, 207,
    210, 213

McClaine, David, 32–33, 136
McDonnell Douglas
  C-17 and fuel inerting system,
    179–80
  design different from Boeing
    fuel tank and air
    conditioning packs
    placement, 230
  MD-83, 19–20
  wiring warning, 57
McDonald, John, 73, 83
McSweeny, Tom, 217, 218, 228,
  232
McWerthor, Ned, 42
Mechanical or electrical cause
  theory, 63, 109, 154–56,
  158–59, 173–73, 184–87,
  190–94, 200–202, 218–21, 236,
  237
Media. *See also* Sgrignoli, Tonice
  and families of victims, 78, 106,
    151–52
  feeding frenzy of, 108–9,
    194–95
  friendly fire theory, reporting
    of, 142
  New York Post reporter Tonice
    Sgrignoli, 105–7, 194–95
  sensational reporting by, 109
  unethical conduct, 209–14
  WNBC reporter John Miller,
    videotaping of crash site,
    51–53
Mershon, Donald, 136, 137
Meyer, Frederick "Fritz," 38,
  131–34, 212
Midland Financial Group, 9,
  188–89
Miller, C. O., 235
Miller, John, 51–53
Miller, Laura Beth, 47–48

Miller, Sam, 41
Missile theory, 54, 57, 62, 63–64,
  65, 109, 117, 132–36, 137–40,
  143, 154, 163–65, 172–73,
  184–87. *See also* Friendly fire or
  U. S. Navy shoot-down theory
Monarch (Coast Guard vessel
  UTL280-501), 43–44
Montoursville
  French teacher and husband, 6,
    29
  High School French Club, 5–6,
    29, 95
  police chief, 169
Morgan, Ron, 138, 140
MSNBC, 108
Murta, Angela, 12–13

New York Air National Guard,
  106th Rescue Wing, 37–39,
  44–46, 131–34, 211–12
New York Air Route Traffic
  Control Center, 137–40
New York Post, 154
Nordstrom, Don, 175–76, 191
North American Aviation, 124, 124n.
Northrop Grumman Corporation,
  148–49
Notes, Godi, 10
Noyes, Mike, 44, 45, 132
NTSB (National Transportation
  Safety Board), 69n, 223. *See also*
  Francis, Robert
  air accidents investigated,
    number of, 223
  and Aviation Disaster Family
    Assistance Act, 88
  and Boeing E-4 study of fuel
    tank heating problems,
    226–27

NTSB, *continued*
    Boeing test of fuel tank
        explosion, 201–2
    Clinton puts agency in charge
        of the families, 171
    EgyptAir 767 crash, 206–7
    and FAA missile concerns,
        137–40
    first responses to crash, 54
    fuel system safety devices
        recommended to FAA, 177,
        180
    glitches and lack of
        communication, 55–56
    and Hall, Jim, 42–43, 86–88, 98,
        137, 143, 172–73, 186,
        197–98, 201, 202, 205–6, 234
    ignition source found, 223–25,
        236, 237
    investigation procedure, 97–98,
        111–12
    investigation team, 59–60,
        91–92
    investigation of ValuJet DC-9
        crash, 55–56, 62, 97, 148,
        162
    lack of equipment, 91–92
    loss of control of Flight 800
        investigation, 56, 60–61,
        97–99m 149–50, 205–6
    mechanical or electrical cause
        investigated, 63, 109,
        184–87, 190–94, 200–202,
        204–5, 218–21, 237
    notified of crash, 42–43
    Office of Aviation Safety,
        147–49
    offices, 147
    out of the loop, 139–40
    previous crashes, investigations,
        147
    recommendations to FAA,
        December, 1996, after crash
        findings, 204–5, 231–32
    Sunset Limited, train
        derailment, Arizona, 97
    test flights, 747–100,
        duplication of Flight 800,
        226–27
    "tin-kickers," 60

Negroni, Christine, x

OBIGGS (onboard inert gas
    generating system), 179–81, 231
O'Connor, John Cardinal, 170
Oelhafen, Kevin, 117–18
O'Flaherty, Johanna, 86, 88–89, 169
O'Meara, Kelly, 116–17, 212–14
Orringer, Paul, 66, 67
Orringer, Shelly, 66

Pan Am corporation
    and Boeing 747 development,
        123, 215
    and fuel tank inerting, 125, 128
Pan Am Flight 103 bombing, 8,
    86, 197
Pan Am Flight 214 crash, 1963,
    65–69, 70, 177, 178, 230, 234
    fuel explosion, 124, 125
    lightning, as ignition of, 68–69,
        126
Pan Am–KLM, Tenerife collision,
    1977, 177–78
Panetta, Leon, 198
Paris Match, 145
Parker Hannifin Company,
    124–28, 176

Passengers on TWA Flight 800. *See also* Bodies
  Alexander, Matthew, 17, 27, 29
  Bellazoug, Myriam, 11
  Buttaroni, Monica Omiccioli and Mirco, 6–7, 18–19, 156–57
  Cremades, Daniel, 17–18, 27, 29, 183–84, 200
  dog, Max, among, 11, 29
  Donaldson, 10
  French Club chaperones, 6, 29
  Gray, Charles Henry, III, 9–10, 83, 169–70, 187–89
  High School French Club, 5–6, 29, 95
  Hill, Susan, 16–17, 27, 29, 234–35
  Hurd, Jamie, 199
  Jenrette, 10
  Johnson, Jed, 10–11
  last seconds of, traumatic injuries to, and deaths of, 27–34, 220–21
  list of, 76, 77, 82, 85, 89, 95
  Loo, Patricia, 12, 27, 61
  Lufkin, 10
  Lychner, Kate, 84–85, 89–90, 200
  Lychner, Pam, 84–85, 89–90, 200
  Lychner, Shannon, 84–85, 89–90, 200
  memorial services for, 170, 236
  Murta, Angela, 12–13
  Notes, Godi, 10
  number of fare-paying, 3
  number of crew and TWA employees, 3
  Rhein, Kurt, 10, 83, 188
  Rose, Katrina, 11
  Story, William, 10
  twins among, 11
  two who switched off flight, 82
  Uzupis, Larissa, 152
  Yee, Judith, 11–13, 23, 27, 29, 61–62, 203–4
Pataki, George, 82, 94, 113–16, 167
Pellinen, Jarl, 40–41
Pena, Frederico, 139
Perfect Storm, The (Junger), 46
Peters, Elmo, 102
Phee, Dan, 41–42
Phillipine Air Lines 737, Manila, 155, 227
Pickard, Tom, 54, 55, 57, 91–92, 144
Pilots
  deadheading, 3
  last actions of Flight 800 pilots, ix, 1–2, 29–30, 220–21
  mandatory retirement, 21
Pirouette search vessel, 111
Poderini, Jean-Claude, 113, 156–57
Port Authority of New York and New Jersey, 77, 81–82
Pressurization technology, 28

Radar control data, and missile theory, 63–64, 137–40, 144–45, 234
Ramada Plaza Hotel, 78, 81, 170–72
Reed, John H., 177
Rhein, Kurt, 10, 83, 188
Richardson, Dennis, 37–39
Roberts, Richard, 84
Rodrigues, Dennis, 162, 202
Romeo (Suffolk County police boat), 75

"Ropes," 21
Rose, John, 11
Rose, Katrina, 11
Russell, Richard, 109, 142, 144–45, 234
Ryan, Mike, 101

Safir, Howard, 95
Salinger, Pierre, 109, 117, 141–45, 209
Sanders, James, 209–12, 233
Sanders, Elizabeth (Liz), 210, 211, 233
Santora, Peter Michael, 61–62, 202–4
Schiliro, Lewis, 206, 207
Schleede, Ron, 56, 137–40, 153, 155
Schoelzel, Hugh, 4
Seebeck, Ken, 40–42, 43–44
Sgrignoli, Tonice, 105–7, 151–53, 194–95
Shanahan, Dennis, 168
Shangvi, Raju, 169–70
Simmons, Olivia, 47
Singing Faith, A (Huber), xiii
60 Minutes, 200–201
Slenski, George, 224–25
Smith, Dennis, 205
Smith, Rick, 45
Snyder, Steven (pilot), ix, 1, 2, 3–4, 7–8, 20–21, 22, 30, 75
Sommers, Mike, 144–45
Soualle, Lydia, 49
Soualle, Patrice, 49–50, 95–96
Spinello, Michael, 86
SST development, 215
Stacey, Terrell, 209–11, 233–34
Static electricity, as ignition source, 63, 192–94

Story, William, 10
Stelzer, Mike, 79–80
Suffolk County Medical Examiner's Office, 48, 49–50. *See also* Bodies
    autopsies, 34, 166–67
    criticism of, by families, 166
    dispute with New York City over families, 93–94
    forensic dentists used by, 168
    Wetli, Dr. Charles, 50–51, 93–94, 108, 166, 167
Sullivan, William "Joe," 178
Sunset Limited, train derailment, Arizona, 97
Swaim, Bob, 111, 155–56, 158–59, 161–62, 186, 190, 218–19, 224–25, 237

Taffert, Ronnie, 203
Tanguay, Leon, 70–71
Termine, Vinnie, 75
Terrorism
    airport security, 7
    Arizona train derailment, 97
    baggage checking, 8
    Colombia, bomb aboard 747, 162
    Federal trial of Yousef, 7, 149, 154
    Pan Am Flight 103 bombing, 8
    theory in Flight 800 crash, 53
Thomas, David, 138
Thomas, Ivor, 126, 127, 181, 191, 202, 226, 228–29, 230, 232
Thompson, Barbara, 204
Thurman, James (Tom), 164
Tomlin, Tara, 188–89
Torche, Melinda, 19, 23
Tower Air, all-747 carrier, 224–25

Trippe, Juan, 122, 125

Tsue, Ted, 125, 176, 219

Tuchman, Gary, 36

Tuchman, Kathy, 36

TWA
  Boeing service bulletin on fuel
    pump wiring, 192
  corporate and financial
    problems, 15, 19–20, 73–74,
    77
  crisis center (Paris), 96
  crisis response team, 48, 85–86,
    96, 105
  criticism of crisis handling by,
    85, 95, 112–13
  disaster response plan, 85–86
  first public statement on crash,
    73, 74–75
  fleet, x, 19–20, 216
  fuel budget, annual, 7
  and fuel tank inerting, 125–26,
    128
  lack of decisions in handling
    families of victims, 78, 83–84
  lack of passenger list, 82–83, 85,
    95
  Paris, Charles de Gaulle
    Airport, 49–50, 95–96
  press conference, day after
    crash, 85, 94
  Snyder, Steven, at, 7–8

TWA DC-9, runway collision,
  1994, 73

TWA Flight 800. See also Black
  boxes; Crash investigation;
  Crash site; Families of
  Passengers; Passengers
  age of plane, 2
  air handling units, 13, 27,
    226–27
  air speed, 23
  altitude at time of explosion, x,
    25, 28
  ascent after explosion, 136–37
  breaking apart of airplane, 1,
    28, 155
  chemical-sniffing dogs tested
    on, 185
  cockpit, 1–2, 23, 24, 28, 30
  conversations between pilots
    and air traffic controllers, ix,
    23
  crew lists and work schedules,
    48
  delay in departure, 7–8, 13, 15
  electrical and/or mechanical
    problems, 24–25, 192,
    220–21
  emergency battery switch, 2
  explosion number 1, 1, 27–30,
    37, 63, 135, 136
  explosion number 2, 1, 31,
    37–38, 63, 135, 136–37
  first class compartment, 29
  flights made by, 3
  flying miles, 3
  fuel tank, center, 3, 27, 28, 63,
    184, 186–87, 193, 224, 225
  inspections of, 2, 3, 216
  number N93119, 2, passim
  passenger manifest, 76, 77, 82
  seating capacity, 3
  take-off, 15–16, 20–21
  time of explosion, ix
  weather, 13, 135
  wiring systems, 216, 220–21,
    237

TWA Flight 800, Rome crash of
  1964, 71, 126

TWA Flight 848 cancellation, 3

TWA Lockheed Constellation
  crash, Milan, Itay, 1959, 69–70

United Flight 585, 1991 crash, 150
USAir, 87
USAir 427, crash at Pittsburgh, 1994, 87, 147
U. S. Air Force
    C-124 Globemaster, 129
    C-141, 129
    E-4 study of fuel tank heating problems, 226–27
    and fuel tank explosion problem, 128–29
    KC-135, 129
    XB-70, 124–25
U. S. S. Grasp, 112, 117–18, 186
Uzupis, Larissa, 152

Valiquette, Joseph, 53, 213–14
ValuJet DC-9 crash, 55–56, 62, 97, 148, 162, 171
Villareale, Tony, 51–52
Vita, Andrew, 205

Walbert, Calvin, 184–85
Wanzenberg, Alan, 11
Weather, 13, 135
Weir, Diana, 134–35, 213
Weiss, Mike, 37
Wetli, Charles, 50–51, 93–94, 108, 166, 167
Whiskey 105E zone, 54–55
Williams, David, 101–4
Wiring
    lack of inspections, 2, 214–19
    and Navy fighter jet accidents, 56–57
    Poly-X, 217, 219

problems on Flight 800, 24–25, 220–21, 223–24
in 747 wing tanks, 191–92
Witnesses to Flight 800 crash
    Baur, Christian, 37–39, 103, 132–33
    Brady, Shaun, 37, 44–46, 133, 134
    Clapp, Chris, 35–36
    FBI questioning of, 133–34
    Fuschetti, Vincent, 32–33
    Johnson, Craig "Jake," 37, 133, 134
    Levine, Naneen, 36, 37
    McClaine, David, 32–33, 136
    Meyer, Frederick "Fritz," 38, 131–34
    Richardson, Dennis, 37–39, 131
    unreliability of memory, 134–37
    Weiss, Mike, 37
Wolk, Arthur, 87
Wright, Bernard, 230

XB-70 (Air Force bomber), 124–25

Yee, Judith, 11–13, 23, 27, 29, 61–62, 203–4
    dog, Max, 11–12, 29
Yee, Ronald, 61–62, 203–4
Yothers, Phillip, 5–6
Yousef, Ramzi Ahmed, 7, 53

Zalosh, Bob, 180